Library

CHEMISTRY

at a Glance

MANSON

PUBLISHING

Copyright © 2005 Manson Publishing Ltd

ISBN: 1–84076–039–7

A CIP catalogue record for this book is available from the British Library.

For full details of all Manson Publishing Ltd titles please write to:
Manson Publishing Ltd, 73 Corringham Road, London NW11 7DL, UK.
Tel: +44(0)20 8905 5150
Fax: +44(0)20 8201 9233
Website: www.mansonpublishing.com

Project manager: Clair Chaventré
Cover design: Cactus Design
Design, illustration, and layout: Cathy Martin, Presspack Computing Ltd

Printed by Replika Press

CONTENTS

CLASSIFYING MATERIALS

SOLIDS, LIQUIDS AND GASES   States of matter

Substances may be classified as solids or liquids or gases.

- Chemistry involves studying substances like water, steel, plastic, oxygen, and salt.
- Substances may be sorted into types by looking for similarities and differences.
- **Solid**, **liquid** and **gas** are the three states of matter.

The universe is believed to be made up of energy and matter. Light, sound, and heat are examples of energy. Substances like rocks, air, and water – which can be touched or felt – are all examples of matter.

Salt
SOLID
Iron   Chalk

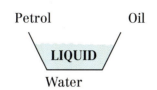
Petrol   Oil
LIQUID
Water

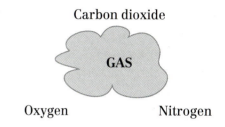
Carbon dioxide
GAS
Oxygen   Nitrogen

The different states have different properties.

SOLID   Has a fixed shape.
Has surfaces all round.

LIQUID   Changes shape to fit its container.
Has one upper surface.

GAS    Spreads and changes shape to fill its container.
Has no real surface.

- In science all materials should be handled safely.
- Particular care has to be taken with some substances.
- Hazard warning signs are used to identify particular problems.    **Harmful**

Questions
1. What are the three states of matter?
2. Name two gases and two liquids.
3. Iron is a typical metal. What is the usual state of most metals?
4. Name a substance which is likely to spread out to fill its container.
5. Will water change shape if it is poured from a bottle into a glass?
6. *Gold, chlorine, helium, diamond, hydrogen, ice, alcohol, and diesel is a list of substances made up of two liquids, three solids, and three gases. Sort them according to their state.*

Density measures how much substance fits into a certain volume (how many grams per cm^3). Generally, gases are least dense and solids are most dense, although some solids float on water.

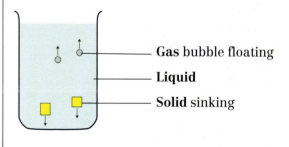
Gas bubble floating
Liquid
Solid sinking

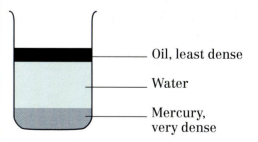
Oil, least dense
Water
Mercury, very dense

Hydrogen, less dense ('lighter') than air

Carbon dioxide, more dense ('heavier') than air

SOLIDS, LIQUIDS AND GASES Changes of state

Many substances can be made to change state by heating or cooling.

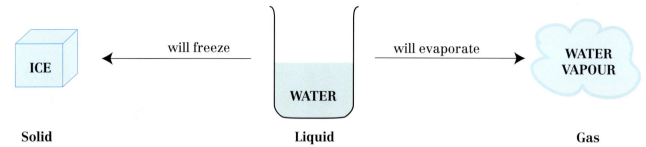

ICE	WATER	WATER VAPOUR
Solid	**Liquid**	**Gas**

will freeze ← → will evaporate

Water is one of the few substances which is commonly found in all three states.

Most metals are solid at room temperature but will become liquid if heated.
Gold above 1064°C   Lead above 328°C

Melting point is that temperature at which a solid begins to melt. In order to remain solid a substance must be kept below its melting point. This is also the temperature at which liquids begin to freeze or solidify.

Substances which are gases can be made liquid if cooled.
Oxygen below –183°C   Butane below –1°C

Boiling point is that temperature at which a liquid begins to boil. In order to remain liquid a substance must be kept above its melting point but below its boiling point. A gas will condense to liquid below its boiling point.

Liquid iron has to be kept above 1535°C or it becomes a solid.

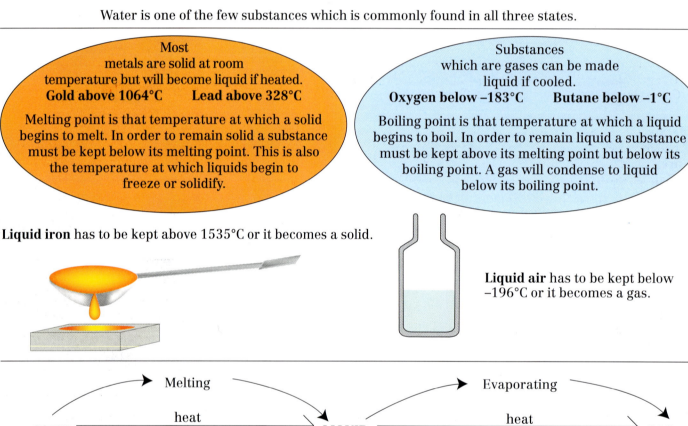

Liquid air has to be kept below –196°C or it becomes a gas.

SOLID — Melting / heat → ← cool / Freezing — LIQUID — Evaporating / heat → ← cool / Condensing — GAS

Solid carbon dioxide has to be kept below –78°C otherwise it turns back into carbon dioxide **gas**.

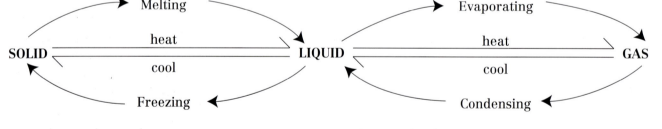

DRY ICE

Solid carbon dioxide → Carbon dioxide gas

Questions
1. What is meant by 'change of state'?
2. What causes substances to change state?
3. What happens to iron above 1535°C?
4. What happens to liquid air at room temperature?
5. What happens to a gas if it condenses?
6. Is it easier to melt iron or gold?
7. The melting point of mercury is –39°C and its boiling point is 357°C. What will be the normal state of mercury at room temperature (25°C)?
8. What will happen to mercury if it is cooled by liquid oxygen?
9. What is dry ice and why is it unusual?

SOLIDS, LIQUIDS AND GASES Particle theory

Differences between solids, liquids and gases can be explained in terms of **particles**.

A **solid** is reluctant to change its shape.

Liquids will follow the shape of their container

Gases want to spread out in all directions

- All substances are made of small particles such as atoms or molecules.
- The properties of a substance depend on how strongly these particles are held in place.
- Gases, liquids and solids behave differently.

The particles are close together in a regular pattern. SOLID	The particles cannot move about but can vibrate.	Solids do not change shape readily because there are strong bonds or forces between the particles which keep them in position.

The particles are close together, but there is no regular pattern. LIQUID	The particles are free to move about beneath the surface of the liquid.	Liquids can change shape because the forces between the particles are not strong enough to stop them moving around.

The particles will spread as far apart as the space allows. GAS	The particles are free to move in all directions.	Gases spread out because the forces between the particles are very weak and cannot keep them together.

- Changes of state occur when substances are heated or cooled.
- Heat gives the **particles** more **energy** to move and overcome the forces between them.
- A solid melts when the particles have enough energy to break free from their normal positions and move around.
- At boiling temperature the particles of a liquid have enough energy to completely break free from each other, and a gas is formed.
- Cooling removes energy from the particles so they slow down.

Questions
1. What are atoms and molecules?
2. Why do solids have a fixed shape?
3. In which state are the forces between particles weakest?
4. In which state are particles close but free enough to move around?
5. Why are changes of state caused by heating or cooling?
6. *Hydrogen is normally a gas but if it is used as a fuel it is carried as a liquid. Why is liquid hydrogen used?*
7. *How is hydrogen made liquid?*

SOLIDS, LIQUIDS AND GASES   Using particles to explain behaviour

The idea of particles is useful for explaining the behaviour of solids, liquids and gases.

- Particle theory explains the properties of substances in terms of **particles**.
- Individual particles like **atoms** and **molecules** are too small to be seen directly.
- Indirect evidence from the behaviour of substances is used to support this theory.

DIFFUSION

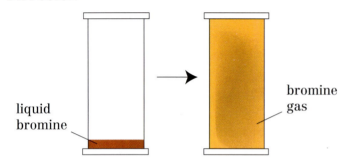

liquid bromine

bromine gas

Observation
The colour or smell of a gas will automatically spread throughout a container or a room.

Explanation
The molecules of the gas (and air) are moving in all directions.

DISSOLVING

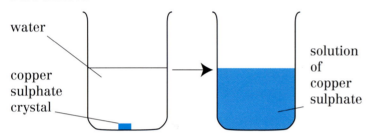

water

copper sulphate crystal

solution of copper sulphate

Observation
A coloured crystal will slowly dissolve and the colour will spread throughout the solution.

Explanation
Molecules of water are free to move beneath the surface and they allow the crystal particles to separate from each other.

PRESSURE

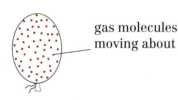

gas molecules moving about

Observation
A balloon filled with air stays inflated as long as the air cannot escape.

Explanation
Molecules of air in the balloon are constantly hitting the sides so that it remains inflated.

PRESSURE

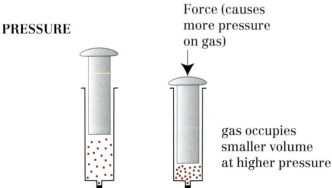

Force (causes more pressure on gas)

gas occupies smaller volume at higher pressure

Observation
If pressure is applied to a gas it will be compressed to a smaller volume.

Explanation
There is a lot of space between the gas molecules and pressure can force them closer together.

- The behaviour of solids, liquids, and gases can be described and explained in terms of particles,
- No one has seen the atoms of a gas such as helium or separate molecules of a liquid like water.
- A single atom of helium is so small that 10,000,000 atoms would fit across 1 mm.
- Each atom of helium has a diameter of 0.0000001 mm.

Questions
1. Why are atoms and molecules not observed directly?
2. Why is it possible for a gas to spread throughout a room? What name is given to this process?
3. When a salt crystal dissolves, why does the taste spread to all parts of the water?
4. Why do gases exert a pressure on the walls of their container?
5. Why can gases be compressed more readily than solids or liquids?

ELEMENTS Metals or non-metals

Elements can be classified as either metals or non-metals.

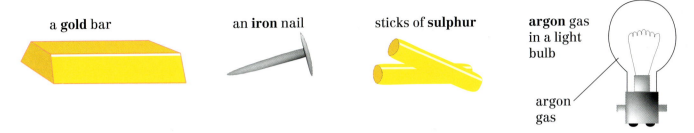

a **gold** bar   an **iron** nail   sticks of **sulphur**   **argon** gas in a light bulb

argon gas

- Air contains a colourless gas called **oxygen** which is essential for respiration.
- Oxygen is a **non-metal**.
- Cooking foil is made from **aluminium**.
- Aluminium is a shiny, reflective **metal** which does not melt in a hot oven.
- By looking at their **properties** elements may be classified as either **metals** or **non-metals**.

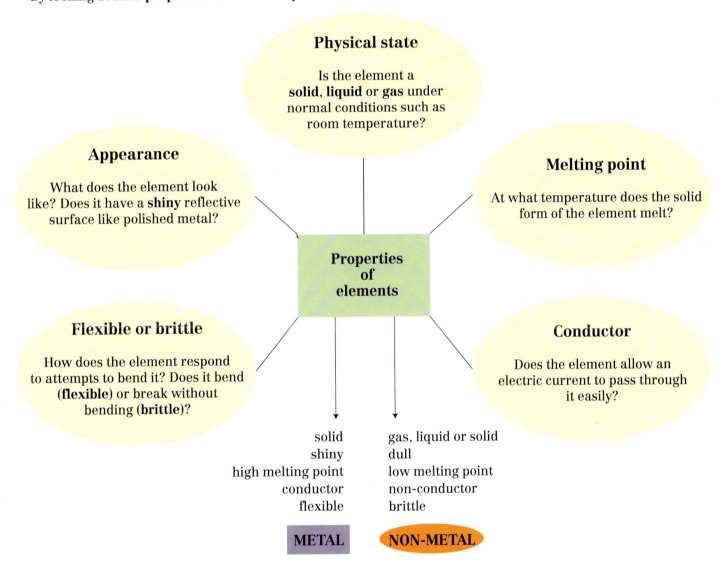

Physical state

Is the element a **solid**, **liquid** or **gas** under normal conditions such as room temperature?

Appearance

What does the element look like? Does it have a **shiny** reflective surface like polished metal?

Melting point

At what temperature does the solid form of the element melt?

Properties of elements

Flexible or brittle

How does the element respond to attempts to bend it? Does it bend (**flexible**) or break without bending (**brittle**)?

Conductor

Does the element allow an electric current to pass through it easily?

solid
shiny
high melting point
conductor
flexible

METAL

gas, liquid or solid
dull
low melting point
non-conductor
brittle

NON-METAL

Questions
1. Name four elements.
2. *Aluminium is used for cooking foil. What does this suggest about its melting point?*
3. Name the gas which is used to fill filament light bulbs.
4. *Why is air not used in light bulbs?*
5. What are the two main types of element?
6. Which type of element can be bent into a new shape?
7. *Name one metal which is not normally found in the solid state,*

ELEMENTS   Classifying elements

Metals and non-metals have different characteristic properties.

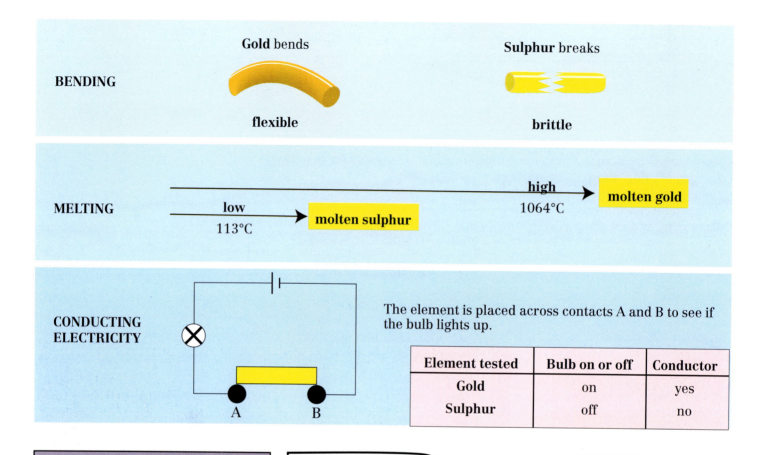

BENDING

Gold bends

flexible

Sulphur breaks

brittle

MELTING

low
113°C

molten sulphur

high
1064°C

molten gold

CONDUCTING ELECTRICITY

The element is placed across contacts A and B to see if the bulb lights up.

A   B

Element tested	Bulb on or off	Conductor
Gold	on	yes
Sulphur	off	no

Gold is a flexible, shiny, yellow solid which melts at 1064°C and conducts electricity.

Iron is a flexible, shiny, silver solid which melts at 1535°C and conducts electricity.

Silicon is a solid which is slightly shiny and melts at 1410°C. It is a **semiconductor** but is not flexible

Argon is a colourless gas. Its melting point is –189°C and it does not conduct electricity.

Sulphur is a brittle, dull, yellow solid. Its melting point is 113°C and it does not conduct electricity.

- Properties such as melting point and conductivity are **physical properties**.
- Oxides of non-metals, for example **sulphur dioxide**, are mostly **acidic**.
- Oxides of metals, for example **calcium oxide**, are mostly **basic**.
- Being **basic** or **acidic** is a difference in **chemical properties**.

Questions
1. Name two elements which conduct electricity.
2. Why did the light stay off when sulphur was tested in the circuit?
3. What would happen if iron were placed in the circuit?
4. Chromium, tungsten, and nickel are all metals. What properties are they likely to have?
5. Phosphorus and iodine are both solid and are non-metals. Will they conduct electricity?
6. Fluorine and chlorine are gases. Are they metals or non-metals?
7. *What is meant by the term semiconductor?*
8. *Why is it difficult to classify silicon as a metal or a non-metal?*
9. *Is silicon a metal or a non-metal?*

ELEMENTS Atoms and atomic symbols

Elements are made up of small particles called atoms.

- Copper **is** an element.
- Helium **is** an element.
- Water **is not** an element.
- Air **is not** an element.
- Elements are the simplest chemicals and cannot be split into simpler substances.
- Each element is made only from itself; copper is only copper, and helium is only helium.

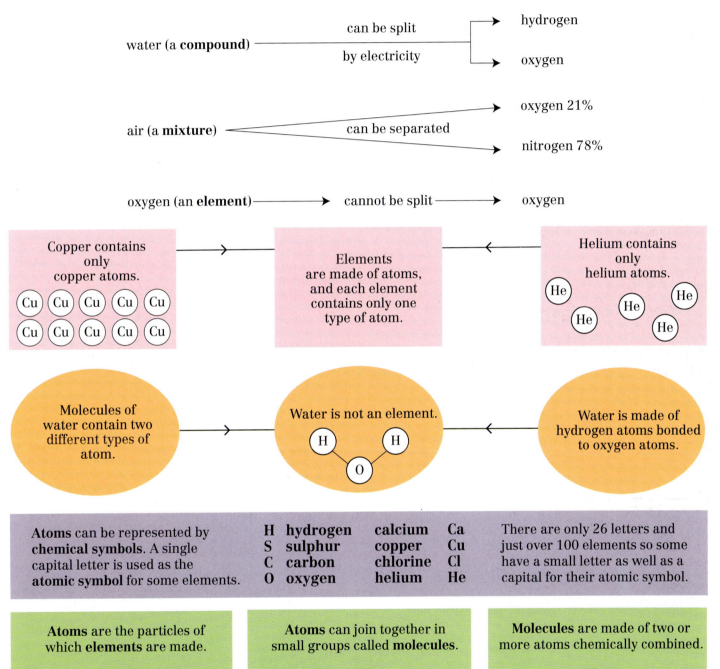

water (a **compound**) —— can be split by electricity —→ hydrogen / oxygen

air (a **mixture**) —— can be separated —→ oxygen 21% / nitrogen 78%

oxygen (an **element**) —→ cannot be split —→ oxygen

Copper contains only copper atoms.

Elements are made of atoms, and each element contains only one type of atom.

Helium contains only helium atoms.

Molecules of water contain two different types of atom.

Water is not an element.

Water is made of hydrogen atoms bonded to oxygen atoms.

Atoms can be represented by **chemical symbols**. A single capital letter is used as the **atomic symbol** for some elements.

H	hydrogen	calcium	Ca
S	sulphur	copper	Cu
C	carbon	chlorine	Cl
O	oxygen	helium	He

There are only 26 letters and just over 100 elements so some have a small letter as well as a capital for their atomic symbol.

Atoms are the particles of which **elements** are made.

He

helium atoms

Atoms can join together in small groups called **molecules**.

H₂
(H–H)
hydrogen molecules

Molecules are made of two or more atoms chemically combined.

H₂O

water molecules

Questions
1. Can copper be split into simpler substances?
2. Can water be split into simpler substances?
3. Name two elements present in air.
4. Which type of particle is present in helium?
5. Why is water not an element?
6. Give the names and atomic symbols of three elements.
7. What is the chemical symbol for calcium and why is it not just C?
8. *Why is there no chemical symbol for air?*

The Periodic Table is a useful way of organising the elements.

There are just over 100 different known elements and they can be organised in many ways.

Alphabetically	By Density (g/cm³)		By Boiling Point (°C)	
Actinium	Osmium	22.5	Helium	−269
Aluminium	Iridium	22.4	Hydrogen	−253
Americium	Platinum	21.5	Neon	−246

- Elements are made up of atoms, which contain even smaller particles called **protons** (plus neutrons and electrons).
- All atoms of the same element have the **same number** of protons.
- Atoms of different elements have **different numbers** of protons.
- The **atomic number** of an element is the number of **protons** in each atom of that element.

atomic number $_6$C $_{13}$Al $_{16}$S $_{79}$Au

How did the Periodic Table develop? **See page 26.**

- The elements can be listed in order of atomic number.
- Elements with similar chemical properties occur at regular intervals (periodically).
- Elements organised using atomic number and chemical properties give a **Periodic Table**.

- The elements are listed in order of atomic number beginning with hydrogen, $_1$H.

  H *He* **Li** Be B C N O F *Ne* **Na** Mg Al Si P S Cl *Ar* **K**

- Elements with similar chemical properties occur at regular intervals, e.g. **Li**, **Na**, **K**; *He*, *Ne*, *Ar*.

In the **Periodic Table** the elements are listed in order of atomic number, **and** elements with similar chemical properties are put in columns called **groups**. This is a **useful way** of organising the chemical elements using atomic number **and** chemical properties.

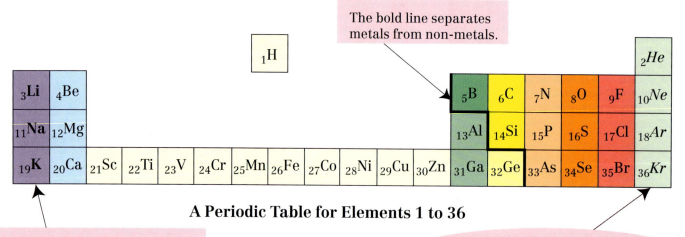

The bold line separates metals from non-metals.

A Periodic Table for Elements 1 to 36

Lithium, sodium, and potassium are all very reactive metals.

Helium, neon, argon, and krypton are all colourless, unreactive gases.

Questions
1. How many known elements are there?
2. Why are certain elements placed in the same column as each other?
3. What are the columns called?
4. What does the bold line between Al and Si indicate?
5. Are there more metals or non-metals?
6. *In the table, oxygen is shown as $_8$O. What does the 8 represent?*
7. Why are He and Ne in the same column?
8. How does the table show that Na, K, and Li have similar properties?

COMPOUNDS Compounds

New substances called **compounds** are formed when elements combine.

Experiment   Heating iron and sulphur
- Iron and sulphur are both **elements**.
- If iron filings and powdered sulphur are mixed together they do not change.
- If a mixture of iron and sulphur is heated the iron and sulphur **combine**.
- When iron and sulphur combine they form a **new substance**.

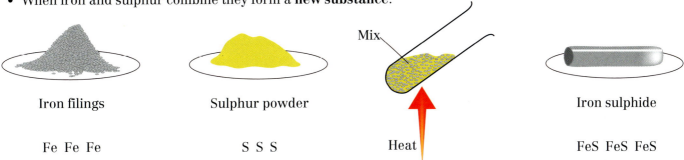

Iron filings	Sulphur powder	Mix / Heat	Iron sulphide
Fe Fe Fe	S S S		FeS FeS FeS

- Iron atoms bond to sulphur atoms to form a new substance called iron sulphide.
- Iron sulphide is a **compound** of iron and sulphur.
- There is a **chemical reaction** during which iron combines with sulphur.

iron   +   sulphur $\longrightarrow$ iron sulphide

Fe   +   S $\longrightarrow$ FeS

When elements combine they form new substances which have different properties.

Experiment   Heating copper in air
- If a piece of copper is heated in air, a black coating forms on the outside of the copper.
- The black substance is a **compound** of copper and oxygen called copper oxide.
- Copper atoms **combine** with oxygen atoms from the air.

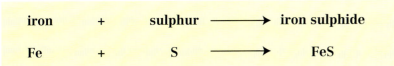

Copper foil   Black coating

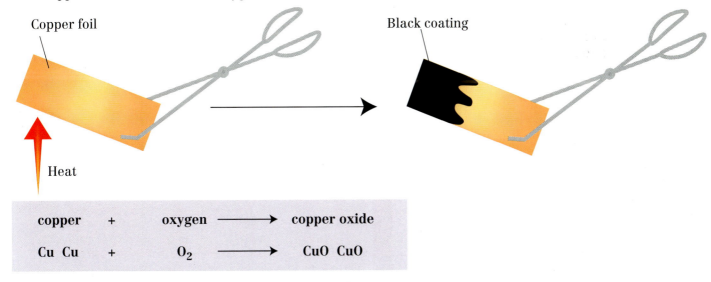

Heat

copper   +   oxygen $\longrightarrow$ copper oxide

Cu Cu   + O_2 $\longrightarrow$ CuO CuO

- Copper oxide and iron sulphide are both compounds.
- Compounds are formed when atoms of different elements form bonds to each other.
- Compounds contain two or more elements which are **chemically combined**.

Questions
1. How can iron be made to combine with sulphur?
2. Name the compound formed when iron combines with sulphur.
3. If a piece of copper is heated in air, what change is seen?
4. What is copper oxide and what does it look like?
5. What keeps the atoms of copper and oxygen together in copper oxide?
6. What are compounds?
7. *Name the compounds formed when copper reacts with sulphur, and iron reacts with oxygen.*
8. *Why do iron and copper not form a compound with each other?*

COMPOUNDS Combination reactions

Combination reactions are those in which elements combine directly to form compounds.

Experiment   Burning sodium in chlorine
- **Combination reactions** often happen if metals and non-metals are heated together.
- If a piece of burning sodium is lowered into chlorine gas, the sodium continues to burn.
- Sodium combines with the chlorine to form a compound called **sodium chloride**.

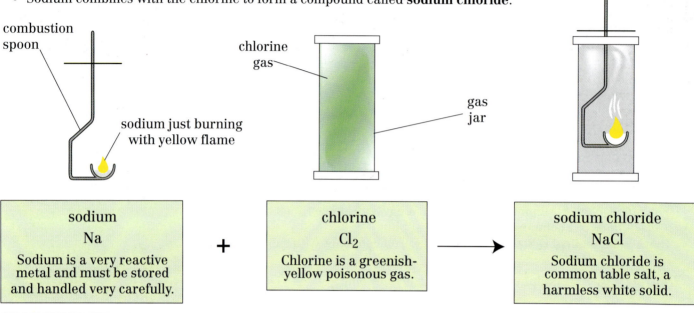

sodium		chlorine		sodium chloride
Na		Cl_2		NaCl
Sodium is a very reactive metal and must be stored and handled very carefully.	**+**	Chlorine is a greenish-yellow poisonous gas.	→	Sodium chloride is common table salt, a harmless white solid.

Experiment   Burning magnesium in air
- Air contains oxygen, which is a reactive gas.
- Many elements react with oxygen if they are heated in air.

Magnesium usually burns
if heated in air.

It combines with oxygen
to form magnesium oxide.

- Magnesium is a silver-grey metal.
- Oxygen is a colourless gas.
- Magnesium oxide is a white powder.

Experiment   Burning sulphur in air and oxygen

Sulphur burns in air.   Sulphur burns in oxygen.

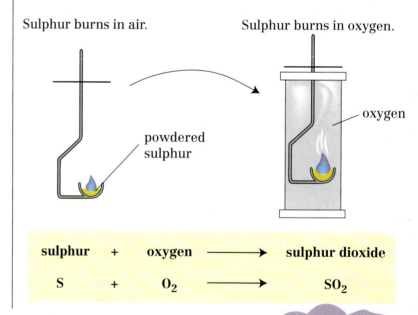

sulphur	+	oxygen	→	sulphur dioxide
S	+	O_2	→	SO_2

- Sulphur is a yellow solid.
- Oxygen is a colourless gas.
- Sulphur dioxide is a poisonous gas.

For
more combination
reactions
see page 70

Questions
1. Which part of the air reacts when magnesium burns?
2. What does magnesium look like after it has burnt? Name the substance formed.
3. What is sodium chloride? In what way is it different from its elements?
4. How does sulphur dioxide differ from oxygen?
5. What is a common method of making elements combine?
6. Why do so many elements react if they are heated in air?

COMPOUNDS Naming chemical compounds

The chemical name given to a compound indicates the elements from which it is made.

A name ending in ...IDE usually indicates compounds containing two elements.

Compounds containing oxygen bonded to one other element are **oxides**	Compounds containing sulphur bonded to one other element are **sulphides**	Compounds containing chlorine bonded to one other element are **chlorides**

Copper oxide
CuO

is a compound of
copper and oxygen

Silver sulphide
Ag_2S

is a compound of
silver and sulphur

Sodium chloride
NaCl

is a compound of
sodium and chlorine

These rules also apply to compounds which seem to have less simple names.

carbon monoxide
CO
⟵ carbon + oxygen ⟶
carbon dioxide
CO_2

Carbon forms two different compounds with oxygen. The names **monoxide** and **dioxide** indicate the number of oxygen atoms bonded to one carbon atom. Similarly, sulphur forms a **dioxide (SO_2)** and a **trioxide (SO_3)**.

A name ending in ...ATE usually indicates compounds with three elements such as a metal, another element, and oxygen.

Compounds containing a metal, carbon, and oxygen are **carbonates**	Compounds containing a metal, sulphur, and oxygen are **sulphates**	Compounds containing a metal, nitrogen, and oxygen are **nitrates**

Calcium carbonate
$CaCO_3$

is a compound of calcium,
carbon, and oxygen

Copper sulphate
$CuSO_4$

is a compound of copper,
sulphur, and oxygen

Sodium nitrate
$NaNO_3$

is a compound of sodium,
nitrogen, and oxygen

Although some formulae look quite complex the endings ...**ide** and ...**ate** still apply.
$AlCl_3$ is aluminium chloride.
$Mg(NO_3)_2$ is magnesium nitrate.
$Al_2(SO_4)_3$ is aluminium sulphate.

Although some names sound unusual they generally indicate the elements in the compound.
Calcium carbide contains calcium and carbon.
Magnesium nitride contains magnesium and nitrogen.
Sodium chlorate contains sodium, chlorine, and oxygen.

Questions
1. Name the elements of which the following substances are compounds:
   CuO   NaCl Ag_2O AgCl Na_2S CO_2 SO_2
2. Name the compounds in question one.
3. *Name compounds which could be formed from the following pairs of elements: Silver and sulphur, sodium and oxygen, copper and chlorine, magnesium and sulphur.*
4. *Name the following compounds: CuS, $CuSO_4$, $CaCO_3$, $CuCO_3$, $MgSO_4$, $MgCO_3$.*

Note: Compounds with the –OH group are called hydroxides. NaOH is sodium hydroxide.

15

COMPOUNDS Chemical formulae

Chemical formulae use atomic symbols to show which elements are present in a compound. Chemical formulae show the proportions of each element present in a compound.

- Atomic symbols represent **single atoms** of an element.
- If atoms are **combined**, small subscript numbers show **how many** atoms are present.
- The **chemical formula** of a substance shows the **type and number** of atoms present.

O	O_2	CO_2
An atom of oxygen	A molecule of oxygen gas	A molecule of carbon dioxide
This is the atomic symbol, not the formula of oxygen gas.	Two atoms are combined. This is the formula of oxygen gas.	Two atoms of oxygen are combined with one atom of carbon.

- The name of a compound and the elements in it can be worked out from its formula.

NaCl **sodium chloride**	Is a compound of sodium and chlorine.	The formula shows one sodium atom and one chlorine atom.
$AlCl_3$ **aluminium chloride**	Is a compound of aluminium and chlorine.	The formula shows one aluminium atom and three chlorine atoms.
CuS **copper sulphide**	Is a compound of copper and sulphur.	The formula shows one copper atom and one sulphur atom.
$CuSO_4$ **copper sulphate**	Is a compound of copper, sulphur, and oxygen.	The formula shows one copper atom, one sulphur, and four oxygen.
$CaCO_3$ **calcium carbonate**	Is a compound of calcium, carbon, and oxygen.	The formula shows one calcium atom, one carbon, and three oxygen.
$NaNO_3$ **sodium nitrate**	Is a compound of sodium, nitrogen, and oxygen.	The formula shows one sodium atom, one nitrogen, and three oxygen.

- Some formulae include brackets.
- Using brackets does not change the name but does affect the number of atoms.
- The number outside a bracket **multiplies** the number of atoms inside.

$NaNO_3 \longrightarrow$
$1 \times Na$
$1 \times N$
$3 \times O$
(one of NO_3)

$Mg(NO_3)_2 \longrightarrow$
$1 \times Mg$
$2 \times N$
$6 \times O$
(two of NO_3)

$(NH_4)_2SO_4 \longrightarrow$
$2 \times N$
$8 \times H$
$1 \times S$
$4 \times O$
(two of NH_4)

Questions
1. Name the elements of which the following substances are compounds: Copper sulphide, copper sulphate, calcium carbonate, sodium carbonate, sodium nitrate, silver nitrate.
2. Name the following compounds: MgO, MgS, $MgSO_4$, $MgCO_3$, $Mg(NO_3)_2$.
3. How many oxygen atoms are used in each of the following: CO_2, $NaNO_3$, $Mg(NO_3)_2$.
4. Name each element and give the number of atoms in $NaCl$, $CuSO_4$, $NaNO_2$, Fe_2O_3, Na_2CO_3, $Ca(NO_3)_2$, $Al_2(SO_4)_3$.

COMPOUNDS Chemical equations

Chemical equations identify the reactants and products of a chemical reaction.

- Chemical equations summarise what happens during a chemical reaction.
- They indicate the starting substances (**reactants**) followed by the substances produced (**products**).
- **Word equations** use chemical names.
- **Symbol equations** use chemical symbols and chemical formulae.

for iron heated with sulphur:

iron	+	sulphur	$\longrightarrow$	iron sulphide
Fe	+	S	$\longrightarrow$	FeS

for carbon burning in air:

carbon	+	oxygen	$\longrightarrow$	carbon dioxide
C	+	O_2	$\longrightarrow$	CO_2

for magnesium metal reacting with chlorine gas:

magnesium	+	chlorine	$\longrightarrow$	magnesium chloride
Mg	+	Cl_2	$\longrightarrow$	$MgCl_2$

for hydrogen reacting with oxygen:

hydrogen	+	oxygen	$\longrightarrow$	water
$2H_2$	+	O_2	$\longrightarrow$	$2H_2O$

Questions

1. Write out and complete the following word equations:
   - magnesium + oxygen $\longrightarrow$; magnesium + sulphur $\longrightarrow$;
   - sodium + chlorine $\longrightarrow$; lead + oxygen $\longrightarrow$.
2. Name the elements which would have to combine to form the following compounds:
   Iron oxide, carbon monoxide, silver chloride, zinc sulphide, hydrogen sulphide, water, sulphur dioxide, magnesium nitride.
3. Write out and complete the following symbol equations:
   - Cu + S $\longrightarrow$; Cu + Cl_2 $\longrightarrow$; S + O_2 $\longrightarrow$;
   - H_2 + S $\longrightarrow$; Zn + S $\longrightarrow$; H_2 + Cl_2 $\longrightarrow$.
4. Suggest a chemical name for water.

MIXTURES Air

In a mixture the components are not chemically combined.

- A burning candle needs oxygen.
- If a gas jar of air is placed over a burning candle, it burns for a while but finally goes out.
- A candle stops burning when there is not enough oxygen left.

> Air is a mixture of gases including **oxygen** (21%) **nitrogen** (78%) and **carbon dioxide** (0.03%)

In **air** the candle continues to burn for a while.

In **oxygen** the candle burns more brightly than in air.

In **carbon dioxide** the candle goes out immediately.

- In **air** the oxygen is not combined with the other gases and is free to react.
- In pure **oxygen** substances burn more brightly and more quickly,
- In air there is less oxygen and other gases get in the way.
- In carbon dioxide the oxygen atoms are already combined with carbon and are not free to react.

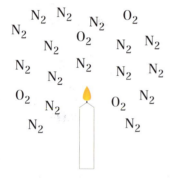

Air contains $1/5$ oxygen

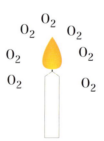

Pure oxygen

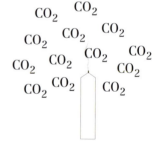

No **free** oxygen

- Oxygen is needed for many reactions such as burning, respiration, and rusting.
- Oxygen is mixed with other gases in the air but mixing does not change its properties.
- The main gas in air is nitrogen, which is generally unreactive.

> The air we breathe out contains less oxygen than the air we breathe in.

> Painting iron or steel protects it from oxygen so that it will not rust.

> Fires can be put out by smothering the flames to keep out oxygen.

Questions
1. Name the two main gases in the air.
2. Why is air described as a mixture rather than a compound?
3. What is the total % of nitrogen and oxygen in the air?
4. Which part of the air is needed for a candle to burn?
5. Why does a candle burn more brightly in oxygen than in air?
6. Why does a candle not burn in carbon dioxide, even though carbon dioxide contains oxygen?
7. *Dry iron does not rust even if air is present. What does this suggest?*
8. What do rusting, burning, and breathing have in common?

MIXTURES Further examples of mixtures

In mixtures the proportions of components may vary.

Gases mix together very easily because their molecules are moving freely. Most gases are colourless and cannot be seen. Air is the most common mixture of gases.

Air.
See page 58

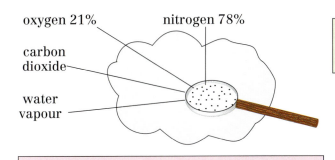

oxygen 21%   nitrogen 78%

carbon dioxide

water vapour

EXHALED AIR — Respiration uses some oxygen so the proportion changes to about 16% oxygen and 4% carbon dioxide.

When the Earth's atmosphere first formed it contained no oxygen. Oxygen was produced by photosynthesis after green plants evolved.

The composition of the **air** does not alter much from day to day except that changes in humidity are caused by changes in amounts of water vapour.

EXHAUST FUMES — Can add gases such as carbon dioxide, carbon monoxide, and nitrogen dioxide to the air.

Water usually contains many dissolved substances. Once dissolved they cannot be seen because the particles spread evenly throughout the solution.

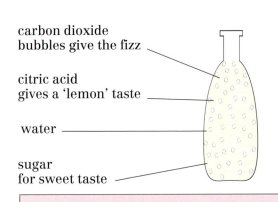

carbon dioxide bubbles give the fizz

citric acid gives a 'lemon' taste

water

sugar for sweet taste

FIZZ — Carbon dioxide gas is not very soluble in water so pressure is used to make it dissolve. Bubbles form as the gas escapes from solution and the drink may go flat.

Different brands of mineral water may contain various dissolved compounds such as sulphates, chlorides and hydrogencarbonates of sodium, magnesium, and calcium.

Lemonade is mostly water with dissolved substances for taste.

TAP WATER — Is not pure but is made fit to drink. Chlorine will have been added to kill bacteria. Hard water will also contain calcium compounds.

Alloys are mixtures of metals with other elements, usually metals, although steels contain iron mixed with small amounts of carbon.

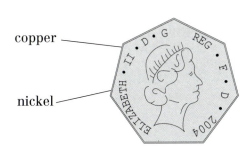

copper

nickel

BRASS — Is an alloy of copper and zinc.

24 carat gold is pure gold and is soft. 9 carat gold is harder and contains 9 parts gold to 15 parts base metal.

'Silver' coins do not contain silver. They are alloys of nickel and copper.

STEEL — The proportion of carbon in steel can vary and this will affect its hardness.

Questions
1. Why are alloys described as mixtures?

2. Name two compounds in mineral water.

MIXTURES   Mixing and separating

Mixing and combining are not the same thing.

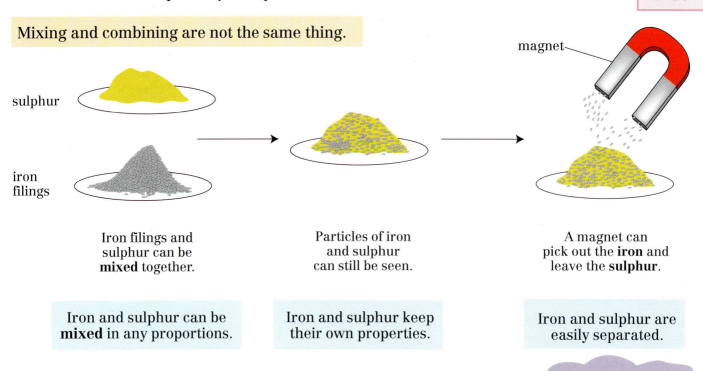

sulphur		magnet
iron filings		

Iron filings and sulphur can be **mixed** together.

Particles of iron and sulphur can still be seen.

A magnet can pick out the **iron** and leave the **sulphur**.

Iron and sulphur can be **mixed** in any proportions.

Iron and sulphur keep their own properties.

Iron and sulphur are easily separated.

- This is a mixture because the iron and sulphur have not combined chemically.
- In most cases it is easier to separate a mixture than to split up a compound.

Iron can combine with sulphur. **See page 13.**

Mixture	Separation	What happens	Why it works
iron/sulphur	use a magnet	Iron is picked out and sulphur is left.	Iron is attracted by a magnet but sulphur is not.
sand/water	filter	Sand stays in the filter paper and water passes through.	Filter paper is porous but the holes are too small for the sand to pass through.
salt/water	heat	Water 'disappears' and salt is left behind.	Heat causes the water to evaporate.
sand/salt	add water	Salt 'disappears' into the water but the sand does not.	Salt is soluble and dissolves in the water but sand is not soluble and does not dissolve.

- In mixtures the components tend to keep their own properties.
- Separation processes make use of differences in physical properties of the components.
- Separation processes do not usually involve chemical reactions.
- A separation might be straightforward or complicated, depending on the type of mixture and which components are needed.

Questions
1. What quantities of iron and sulphur are needed to make a mixture?
2. By looking at a mixture of iron and sulphur, how is it possible to tell that neither has been changed by the mixing process?
3. Why does a magnet separate iron from a mixture of iron and sulphur but not from a compound of iron and sulphur?
4. *If chalk is mixed with water and the mixture filtered, what happens?*
5. *What happens to salt when it is mixed with water, and how can the salt be recovered?*
6. What happens when a mixture of salt and sand is added to water and the mixture stirred? *How can the sand be separated and collected? How can the salt be collected?*

MIXTURES Separation processes

Mixtures are usually separated by physical processes rather than chemical reactions.

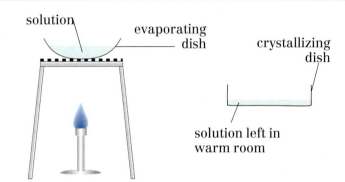

Process	Filtration.
Uses	To separate solids from liquids.
Example	Sand and water.
How it works	Small holes in the paper let only the liquid through.
Limitations	Very fine particles of solid might get through the holes or block them. It does not work if the solid is dissolved.

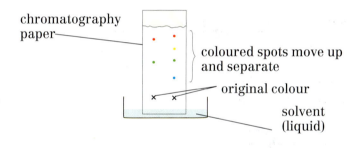

Process	Evaporation.
Uses	To extract solids from solution.
Example	Salt from water.
How it works	The liquid evaporates and the solid is left behind.
Limitations	If crystals are required the evaporation must be slow.

Process	Chromatography.
Uses	To separate mixtures of coloured substances.
Example	Separating coloured inks or food colourings.
How it works	A liquid carries the different colours at different speeds across the surface of the chromatography paper.

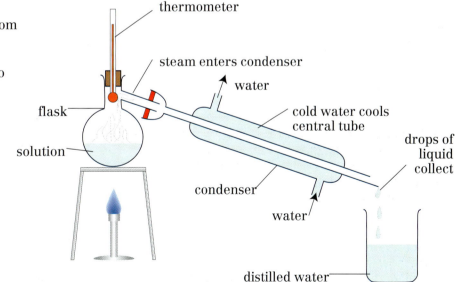

Process	Distillation.
Uses	To recover the liquid from a solution.
Example	Water from salt water.
How it works	The solution is boiled so that steam is given off. The steam is cooled and condenses to re-form water.
Limitations	Works best for solutions of solid in liquid.

Fractional distillation.
See page 45

See page 45

Questions
1. *Give two examples of mixtures which could be made clear by filtering.*
2. Why does filtration not separate salt from sea water?
3. How may pure water be obtained from sea water?
4. If a dish of sea water is left in a warm room, what happens?
5. *Why is drinking water not purified by distillation?*

REVIEW QUESTIONS Classifying materials I

1. Water was kept at –10°C and then slowly heated until it reached 110°C. Describe and name the changes of state which occur between –10°C and 110°C, and indicate the temperatures at which these changes take place.

2. Draw diagrams to show the arrangements of particles in a typical solid, liquid, and gas, and describe the movement of the particles in each state.

3. Describe the processes of dissolving and diffusion, and explain each one in terms of particles.

4. (a) List the main properties which distinguish metals from non-metals.
    (b) Mercury is a liquid at normal temperatures so why is it classified as a metal?

5. (a) Why is hydrogen chosen as the first element in the Periodic Table?
    (b) Why are sodium and potassium in the same column of the Periodic Table?
    (c) What name is given to the columns in the Periodic Table?
    (d) One column comprises colourless, unreactive gases. Name two of these gases.
    (e) Which type of element is most common in the Periodic Table?

6. (a) From the table of melting points and boiling points classify the elements A to F as solid, liquid or or gas at a temperature of 20°C.
    (b) Which, if any, would undergo a change of state if the temperature rose to 2000°C?
    (c) Which would not change state even if cooled to absolute zero (–273°C) from 20°C?
    (d) Which one is water?
    (e) Which one is tungsten, the metal used for making filaments in light bulbs?
    (f) Which one would condense first as the temperature fell from 20°C?

Element	A	B	C	D	E	F
Melting point °C	–182	–117	801	–201	0	3410
Boiling point °C	–164	78	1413	–196	100	5660

7. The diagrams show the particles of different substances. Sort them into elements, mixtures, and compounds.

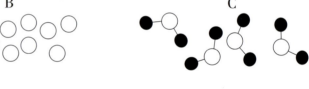

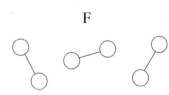

8. Using iron and sulphur as examples, describe and explain the differences between elements, mixtures and compounds.

9. Describe the evidence which suggests that the oxygen in the air is mixed and not combined with the other gases.

10. With the aid of diagrams describe how a mixture of salt and sand can be separated in order to obtain samples of salt crystals and clean, dry sand.

11. Explain the difference between evaporating salt water and distilling salt water.

12. From the formulae listed state the numbers and types of atoms in each compound, and name each compound:

FeS   NaCl CO_2 $FeSO_4$ $CaCO_3$ Na_2CO_3 $Ca(OH)_2$ $Al_2(SO_4)_3$.

ATOMIC STRUCTURE Protons, neutrons, and electrons
Atomic number and mass number

An element is a substance containing only one type of atom.
Atoms consist of a nucleus and electrons that orbit around
the nucleus.
All atoms of a particular element have the same number
of protons.
Atoms of different elements have different numbers
of protons.

Particle in the atom	Relative mass	Relative charge
proton	1	+1
neutron	1	0
electron	negligible	−1

Sodium atom

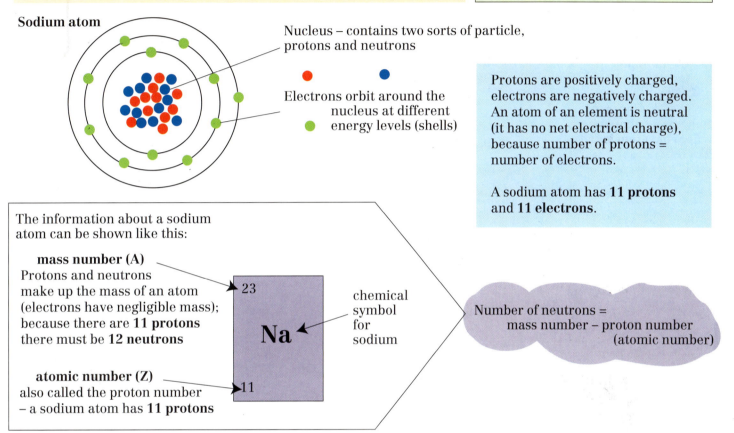

Nucleus – contains two sorts of particle, protons and neutrons

Electrons orbit around the nucleus at different energy levels (shells)

Protons are positively charged, electrons are negatively charged. An atom of an element is neutral (it has no net electrical charge), because number of protons = number of electrons.

A sodium atom has **11 protons** and **11 electrons**.

The information about a sodium atom can be shown like this:

mass number (A)
Protons and neutrons make up the mass of an atom (electrons have negligible mass); because there are **11 protons** there must be **12 neutrons**

23

Na

11

chemical symbol for sodium

Number of neutrons =
    mass number – proton number
                (atomic number)

atomic number (Z)
also called the proton number
– a sodium atom has **11 protons**

Another way of showing a sodium atom is like this

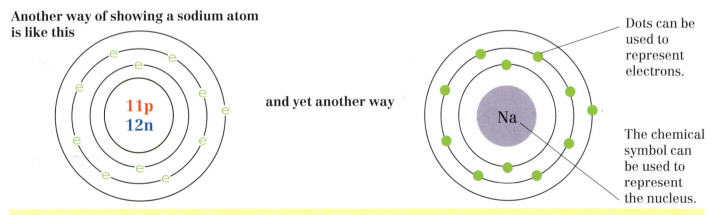

11p
12n

and yet another way

Na

Dots can be used to represent electrons.

The chemical symbol can be used to represent the nucleus.

Questions
1. How many types of atom are there in an element?
2. Which particles are found in the nucleus?
3. What information about an atom is given by the atomic number of an element?
4. What information about an atom is given by the mass number?
5. Which particles in an atom contribute to the mass of an atom?
6. Which particles in an atom are charged?
7. Why are atoms of elements described as being neutral?

ATOMIC STRUCTURE Electron arrangements

Atoms of different elements have different numbers of protons and, therefore, different atomic numbers.

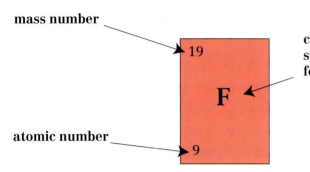

mass number

chemical symbol for fluorine

atomic number

Particles in a fluorine atom

atomic number = 9 so there are **9 protons** and **9 electrons**

mass number = 19 so there are **10 neutrons**

One way of showing a fluorine atom is like this

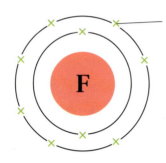

Crosses can be used to represent electrons.

Electron arrangements

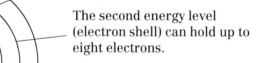

nucleus

The first energy level (electron shell) can hold up to two electrons.

The second energy level (electron shell) can hold up to eight electrons.

The third energy level (electron shell) has a stable arrangement when its holds eight electrons, although it can hold a total of 18.

Electrons orbit around the nucleus at different energy levels (sometimes called electron shells).

Each electron in the atom is at a particular energy level (in a particular shell).

Electrons in an atom occupy the lowest available energy levels (innermost electron shells).

When an energy level is full, further electrons orbit at the next level.

Questions
1. Atoms of elements contain protons, neutrons, and electrons. How do atoms of different elements differ from one another?
2. How many electrons can orbit a nucleus at:
   (a) the first energy level
   (b) the second energy level
   (c) the third energy level?
3. *What would be the electron arrangements for elements with the atomic numbers:*
   (a) 17   (b) 19   (c) 13?
4. *For each of the following elements:*
   (a) give the number of each type of particle
   (b) give the electron arrangements
   (c) draw a diagram to show the structure of an atom.

31
P
15

24
Mg
12

Sodium atom
Notice that there are:
  2 electrons at the first energy level – it is full;
  8 electrons at the second energy level – it is full;
  1 electron at the third energy level – it is not full.

The electron arrangement can be written as 2.8.1

Fluorine atom
Notice that there are:
  2 electrons at the first energy level – it is full;
  7 electrons at the second energy level – it is not full.

The electron arrangement can be written as 2.7

Reactions of elements depend on the arrangement of electrons. Elements with the same number of electrons in the outer energy level react in a similar way.

ATOMIC STRUCTURE Isotopes

Isotopes are atoms of the same element with the same number of protons, but different numbers of neutrons.
Isotopes have the same atomic number, but different mass numbers.
Isotopes of the same element have the same chemical properties, although there will be very small differences in their physical properties.
Most elements have more than one isotope.
Some isotopes are radioactive.

Hydrogen

1		
H	1p	1e
1		

(a hydrogen atom has no neutrons)

Most hydrogen atoms are like this.

2			
H	1p	1n	1e
1			

This isotope is called **deuterium**.

3			
H	1p	2n	1e
1			

This isotope is called **tritium**.

All three isotopes of hydrogen have the same atomic number (1) – they all have one proton, but they have different numbers of neutrons and so have different mass numbers.

Deuterium is used as a moderator in nuclear reactors.

Tritium has a possible use in producing energy by nuclear fusion.

Carbon

12			
C	6p	6n	6e
6			

Most carbon atoms are like this.

This isotope is called carbon-12

13			
C	6p	7n	6e
6			

A few carbon atoms are like this.

This istope is called carbon-13

14			
C	6p	8n	6e
6			

Some carbon atoms are like this.
This isotope is called carbon-14

All three isotopes of carbon have the same atomic number (6) – all have 6 protons, but they have different numbers of neutrons and so have different mass numbers.

$^{14}_{6}C$ is used in radiocarbon dating. This isotope is present in constant small quantities in the atmosphere. Once an animal or plant has died the only change in the $^{14}_{6}C$ content is due to radioactive decay. By knowing the rate of decay ($^{14}_{6}C$ has a half life of 5730 years) and the level of radioactivity in the specimen, its age can be calculated. It is one of the methods used by archaeologists to determine the age of organic remains.

Chlorine

35			
Cl	17p	18n	17e
17			

75% of chlorine atoms are like this.

37			
Cl	17p	20n	17e
17			

25% of chlorine atoms are like this.

Both isotopes of chlorine have the same atomic number (17) – they both have 17 protons, but they have different numbers of neutrons and so have different mass numbers.

Sometimes a Periodic Table will give chlorine a relative atomic mass of 35.5. This takes account of the proportions of the two isotopes.

Questions
1. Why are $^{12}_{6}G$ $^{13}_{6}C$ and $^{14}_{6}C$ all considered to be atoms of the same element?
2. Name the three isotopes of hydrogen and write symbols for them.
3. From the examples on this page give two radioactive isotopes.
4. Explain why a Periodic Table might give chlorine relative atomic mass of 35.5.
5. Two isotopes of uranium are $^{235}_{92}U$ and $^{238}_{92}U$. For each of these isotopes give
   (a) the atomic number   (b) the mass number   (c) the number of protons
   (d) the number of neutrons   (e) the number of electrons.

PERIODIC TABLE 1   Development of ideas

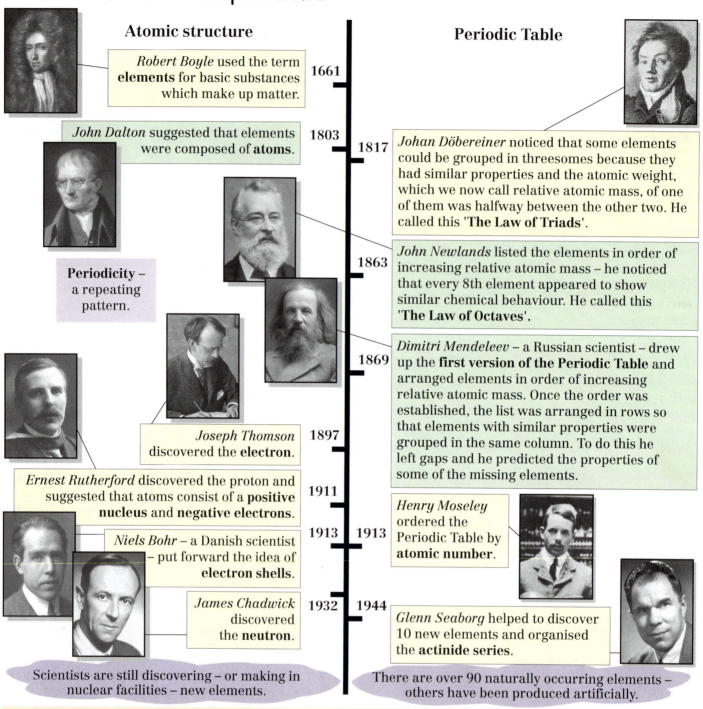

Atomic structure

Robert Boyle used the term **elements** for basic substances which make up matter.
1661

John Dalton suggested that elements were composed of **atoms**.
1803

Periodicity – a repeating pattern.

Joseph Thomson discovered the **electron**.
1897

Ernest Rutherford discovered the proton and suggested that atoms consist of a **positive nucleus** and **negative electrons**.
1911

Niels Bohr – a Danish scientist – put forward the idea of **electron shells**.
1913

James Chadwick discovered the **neutron**.
1932

Scientists are still discovering – or making in nuclear facilities – new elements.

Periodic Table

1817
Johan Döbereiner noticed that some elements could be grouped in threesomes because they had similar properties and the atomic weight, which we now call relative atomic mass, of one of them was halfway between the other two. He called this 'The Law of Triads'.

1863
John Newlands listed the elements in order of increasing relative atomic mass – he noticed that every 8th element appeared to show similar chemical behaviour. He called this 'The Law of Octaves'.

1869
Dimitri Mendeleev – a Russian scientist – drew up the **first version of the Periodic Table** and arranged elements in order of increasing relative atomic mass. Once the order was established, the list was arranged in rows so that elements with similar properties were grouped in the same column. To do this he left gaps and he predicted the properties of some of the missing elements.

1913
Henry Moseley ordered the Periodic Table by **atomic number**.

1944
Glenn Seaborg helped to discover 10 new elements and organised the **actinide series**.

There are over 90 naturally occurring elements – others have been produced artificially.

Scientists regarded the Periodic Table first as a curiosity and later as a scientific tool. Now it is regarded as an important summary of the structure of atoms.

Originally the elements were arranged in order of increasing relative atomic mass. Now they are arranged in order of increasing atomic number, as evidence has become available to show that the chemical properties of elements vary according to atomic number.
When the Periodic Table was arranged in order of relative atomic mass argon, $^{40}_{18}$Ar, was placed after potassium, $^{39}_{19}$K.

Questions
1. What contribution did each of the following scientists make to our understanding of the structure of the atom:
   (a) Joseph Thomson   (b) Ernest Rutherford
   (c) Niels Bohr   (d) James Chadwick?
2. What were
   (a) 'The Law of Triads' (b) 'The Law of Octaves'?
3. How did Mendeleev arrange the elements in his version of the periodic table?
4. When Mendeleev drew up the first periodic table, why did he leave gaps?
5. Why do you think the elements were not arranged in order of increasing atomic number until 1913?
6. How many naturally occurring elements are there?
7. *Approximately how many elements have been discovered?*

PERIODIC TABLE I The first 20 elements

On this page mass numbers of elements have been omitted.

Group 1	2											3	4	5	6	7	0
					H Hydrogen 1												He Helium 2
Li Lithium 3	Be Beryllium 4											B Boron 5	C Carbon 6	N Nitrogen 7	O Oxygen 8	F Fluorine 9	Ne Neon 10
Na Sodium 11	Mg Magnesium 12											Al Aluminium 13	Si Silicon 14	P Phosphorus 15	S Sulphur 16	Cl Chlorine 17	Ar Argon 18
K Potassium 19	Ca Calcium 20	Sc Scandium 21	Ti Titanium 22	V Vanadium 23	Cr Chromium 24	Mn Manganese 25	Fe Iron 26	Co Cobalt 27	Ni Nickel 28	Cu Copper 29	Zn Zinc 30	Ga Gallium 31	Ge Germanium 32	As Arsenic 33	Se Selenium 34	Br Bromine 35	Kr Krypton 36
Rb Rubidium 37	Sr Strontium 38	Y Yttrium 39	Zr Zirconium 40	Nb Niobium 41	Mo Molybdenum 42	Tc Technetium 43	Ru Ruthenium 44	Rh Rhodium 45	Pd Palladium 46	Ag Silver 47	Cd Cadmium 48	In Indium 49	Sn Tin 50	Sb Antimony 51	Te Tellurium 52	I Iodine 53	Xe Xenon 54

Period (1–5 down the right-hand side)

Vertical columns are called **Groups**. The Group number is the same as the number of electrons in the outer energy level (electron shell).
Horizontal rows are called **Periods**. The Period number is the same as the energy level (electron shell) being filled.
In Period 2, the second energy level (electron shell) is gradually filled up with electrons.
In Period 3, electrons are going into the third energy level (electron shell).

Activity

Copy and complete the following table.

Name of element	Chemical symbol	Atomic number	Number of protons	Number of electrons	Electron arrangement
hydrogen	H	1		1	1
helium	He	2	2		2
	Li		3	3	2.1
beryllium		4	4	4	
boron	B	5		5	2.3
	C	6	6		2.4
nitrogen		7		7	
	O		8	8	2.6
fluorine	F	9		9	2.7
neon		10	10		
	Na	11		11	2.8.1
magnesium		12	12	12	
aluminium	Al		13		2.8.3
	Si	14		14	
phosphorus	P		15	15	2.8.5
	S	16	16		
chlorine		17		17	2.8.7
argon	Ar		18		2.8.8
potassium		19		19	
	Ca	20	20		2.8.8.2

Questions

1. (a) Which elements have stable outer energy levels of electrons?
   (b) To which Group of the Periodic Table do they belong?
2. List in order the elements in Period 3.
3. In Period 3, which energy level is gradually being filled up with electrons?
4. (a) How many full energy levels of electrons are there in a magnesium atom?
   (b) How many electrons are there in the outer energy level of a calcium atom?
5. (a) What are the electron arrangements of oxygen and sulphur?
   (b) To which group of the Periodic Table do they belong?

PERIODIC TABLE 1   Patterns of electron arrangements

Elements in the same **Group** have similar **physical** and **chemical properties**.
Similarities between reactions of elements in the same Group can be explained by the number of electrons in the outer energy level (electron shell).

On this page mass numbers of elements have been omitted.

1	2											3	4	5	6	7	0
						H Hydrogen 1											He Helium 2
Li Lithium 3	Be Beryllium 4											B Boron 5	C Carbon 6	N Nitrogen 7	O Oxygen 8	F Fluorine 9	Ne Neon 10
Na Sodium 11	Mg Magnesium 12											Al Aluminium 13	Si Silicon 14	P Phosphorus 15	S Sulphur 16	Cl Chlorine 17	Ar Argon 18
K Potassium 19	Ca Calcium 20	Sc Scandium 21	Ti Titanium 22	V Vanadium 23	Cr Chromium 24	Mn Manganese 25	Fe Iron 26	Co Cobalt 27	Ni Nickel 28	Cu Copper 29	Zn Zinc 30	Ga Gallium 31	Ge Germanium 32	As Arsenic 33	Se Selenium 34	Br Bromine 35	Kr Krypton 36
Rb Rubidium 37	Sr Strontium 38	Y Yttrium 39	Zr Zirconium 40	Nb Niobium 41	Mo Molybdenum 42	Tc Technetium 43	Ru Ruthenium 44	Rh Rhodium 45	Pd Palladium 46	Ag Silver 47	Cd Cadmium 48	In Indium 49	Sn Tin 50	Sb Antimony 51	Te Tellurium 52	I Iodine 53	Xe Xenon 54
Cs Caesium 55	Ba Barium 56	La Lanthanum 57	Hf Hafnium 72	Ta Tantalum 73	W Tungsten 74	Re Rhenium 75	Os Osmium 76	Ir Iridium 77	Pt Platinum 78	Au Gold 79	Hg Mercury 80	Tl Thallium 81	Pb Lead 82	Bi Bismuth 83	Po Polonium 84	At Astatine 85	Rn Radon 86
Fr Francium 87	Ra Radium 88	Ac Actinium 89															

Note 14 elements – the 'lanthanides' – have been omitted from the table between La and Hf.

Group 1 **Alkali metals**	**Group 2** **Alkaline earth metals**	**Group 7** **Halogens**	**Group 0** **Noble gases**
Li 2.1 Na 2.8.1 K 2.8.8.1	Mg 2.8.2 Ca 2.8.8.2	F 2.7 Cl 2.8.7 Br 2.8.18.7	He 2 Ne 2.8 Ar 2.8.8
These **elements react in a similar way** because **they all have one electron** in the outer energy level, e.g. they all react vigorously with water.	These elements **react in a similar way** because **they both have two electrons** in the outer energy level.	These **elements react in a similar way** because **they all have seven electrons** in the outer energy level, e.g. they all react with metals to give ionic compounds.	These elements are **unreactive** because they have a **stable arrangement of outer electrons**.

Questions

1. (a) What are the electron arrangements of lithium, sodium, and potassium?
   (b) Where are these elements in the Periodic Table?
2. (a) What are the electron arrangements of beryllium, magnesium, and calcium?
   (b) Where are these elements in the Periodic Table?
3. (a) What are the electron arrangements of fluorine and chlorine?
   (b) Where are these elements in the Periodic Table?
4. (a) What are the electron arrangements of helium, neon, and argon?
   (b) Where are these elements in the Periodic Table?
5. *What is the relationship between the number of electrons in the outer energy level and the Group number in the Periodic Table?*
6. *How many electrons are there in the outer energy levels of:*
   (a) *rubidium*   (b) *strontium*   (c) *bromine*   (d) *iodine*
   (e) *krypton*   (f) *barium*   (g) *xenon*   (h) *caesium?*

IONS AND BONDING Outer electrons

In order to react atoms must form chemical bonds to other atoms.
Electrons in the outer energy level (outer shell) are used to form bonds.

Fluorine	$_9F$	2 . 7	Group 7	a **reactive non-metal**
Neon	$_{10}Ne$	2 . 8	Group 0	an **unreactive gas**
Sodium	$_{11}Na$	2 . 8 . 1	Group 1	a **reactive metal**

- Fluorine and sodium form compounds with other elements.
- Atoms of fluorine and sodium are able to form chemical bonds to other atoms.
- Neon atoms do not form bonds, which means that **neon** is completely **unreactive**.

F_2
Fluorine forms a molecule of two combined fluorine atoms.

Na_2O
Sodium oxide is a compound formed when sodium reacts with oxygen.

NaF
Sodium fluoride is a compound in which sodium is combined with fluorine.

Ne
Neon gas contains separate atoms of neon. It does not combine with other elements.

- Neon has a complete shell of eight outer electrons.
- Neon is unreactive because it has a stable arrangement of outer electrons.
- The Noble Gases all have stable arrangements of outer electrons and are unlikely to react.

Noble Gases?
See page 91.

$_2He$	2	outer energy level full	Stable arrangement of outer electrons	–	unreactive
$_{10}Ne$	2 . 8	outer energy level full	Stable arrangement of outer electrons	–	unreactive
$_{18}Ar$	2 . 8 . 8	eight outer electrons	Stable arrangement of outer electrons	–	unreactive

- The chemical reactions of an atom depend on the number of outer electrons.
- Sodium and potassium are reactive metals, and both have only one outer electron.
- Chlorine and fluorine have seven outer electrons, and are both reactive gases.

More information?
See page 89.

- The reaction between atoms of sodium and chlorine involves electrons in the outer shell.
- Outer electrons are used to form bonds and this determines how atoms react.
- Argon does not change its outer electrons therefore cannot form bonds and does not react.

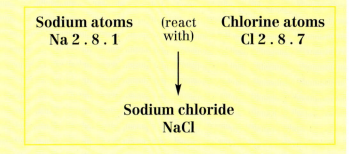

Sodium atoms   (react   **Chlorine atoms**
Na 2 . 8 . 1   with)   Cl 2 . 8 . 7

Sodium chloride
NaCl

Questions

1. What is special about the outer electron arrangement of neon?
2. Why does neon not form compounds?
3. Why are the elements helium and argon unreactive?
4. Which electrons are used to form chemical bonds?
5. Why is potassium reactive and why is chlorine reactive?
6. *Write the electron arrangement of sulphur, $_{16}S$ and predict if it will be reactive.*
7. *Why does $_3Li$ react with $_9F$?*

IONS AND BONDING Ions I

Ions are formed when atoms **gain** or **lose electrons**. Some atoms react by forming ions,

- If a chlorine atom gains an electron its outer energy level becomes stable.
- Chlorine reacts in order to gain an outer electron.

$$Cl \; 2.8.7 \qquad + \qquad e^- \qquad \rightarrow \qquad Cl^- \; 2.8.8$$

chlorine atom   electron   chloride ion

> Cl is the symbol for a chlorine atom, which has equal numbers of protons and electrons.

> Cl⁻ is the symbol for a chloride ion. The negative (⁻) symbol indicates that there is one extra electron.

- A sodium atom has one more outer electron than it needs for a stable outer shell.
- Sodium atoms react in order to lose the outer electron.

$$Na \; 2.8.1 \qquad + \qquad e^- \qquad \rightarrow \qquad Na^+ \; 2.8$$

sodium atom   electron   sodium ion

> The sodium atom has 11 protons (+) and 11 electrons (–).

> The sodium ion has 11 protons (+) but only 10 electrons (–).

- Many atoms can gain or lose electrons to obtain a stable arrangement of **outer electrons**.
- They will attain the electron arrangement of the **nearest Noble Gas**.
- This is what happens during some chemical reactions.

> Only outer electrons are involved in chemical reactions.

e.g.   K 2.8.8.1   →   K⁺ 2.8.8   potassium ion

                                Ar 2.8.8   argon atom

    Cl 2.8.7   →   Cl⁻ 2.8.8   chloride ion

- Atoms form ions by reacting with other atoms according to certain rules.
- Atoms will gain or lose enough electrons to achieve a **stable outer energy level**.
- The **charge** on an ion indicates the number of electrons **gained** or **lost**.
- Metals usually lose electrons to form **positive ions (cations)**.
- Non-metals usually gain electrons to form **negative ions (anions)**.

Ions are **charged** particles made from atoms, or groups of atoms bonded together.

Questions

1. How does a chlorine atom change its outer energy level to eight electrons?
2. What is the difference between a chlorine atom and chloride ion?
3. Why does the chloride ion (Cl⁻) have a negative charge?
4. Why does a sodium atom lose one electron when it reacts?
5. Give the symbol and name the particle formed when sodium loses an electron.

6. Why do certain atoms react?
7. *What determines whether an atom loses or gains electrons?*
8. *Which of the following will form positive ions if they react:*

   Na, Al, Cl, O, Mg, Cu?

9. *Explain why argon does not form ions and give another example.*

IONS AND BONDING Ions II

In some cases the Periodic Table can be used to predict how many electrons an atom is likely to gain or lose during a chemical reaction.

- Elements in Group 7 all have seven outer electrons.
- The atoms of elements in Group 7 can react by gaining an outer electron to form a negative ion.

1	2											3	4	5	6	7	0
							H										
Li^+															O^{2-}	F^-	
Na^+	Mg^+											Al^{3+}			S^{2-}	Cl^-	
K^+	Ca^+															Br^-	
Rb^+																I^-	
Cs^+																	

Ions and their charge

Group 1 metals **lose one** outer electron	K 2.8.8.1 $\rightarrow$	K^+ 2.8.8
Group 2 metals **lose two** outer electrons	Mg 2.8.2 $\rightarrow$	Mg^{2+} 2.8
Oxygen atoms – from Group 6 – **gain two** electrons	O 2.6 $\rightarrow$	O^{2-} 2.8

For some elements it is difficult to predict the charge on the ion formed, but the rules about metals and non-metals still apply.

Iron and copper are metals and form positive ions:

$$Fe^{2+} \quad Fe^{3+} \quad Cu^{2+}$$

Nitrogen and phosphorus can form simple negative ions – N^{3-} (nitride) and P^{3-} (phosphide) – but these are not very common and these elements are more often found in ions such as NO_3^- (nitrate) and PO_4^{3-} (phosphate).

Questions

1. If a metal and a non-metal react, which will form the + ion and which the – ion?
2. Chlorine, bromine, and iodine are all in Group 7 of the Periodic Table. Write symbols for the chloride, bromide, and iodide ions.
3. Why do lithium, sodium, and potassium all form ions with a single + charge?
4. *The change in electrons as ions form can be shown as, for example, Mg 2.8.2 $\rightarrow$ Mg^{2+} 2.8. Show the same changes when the atoms listed form ions:*

   $_3Li$, $_{13}Al$, $_{20}Ca$, $_9F$, $_{16}S$.

5. *Why is it difficult to predict the ions formed by $_6C$?*
6. *Why are the elements Sc to Zn all likely to form positive ions?*
7. *Name the following ions:*

   Cl^-, Br^-, O^{2-}, S^{2-}, SO_4^{2-}, CO_3^{2-}.

IONS AND BONDING Ionic bonds

An ionic bond is a strong electrostatic attraction between oppositely charged ions.

- Ions are formed during certain chemical reactions.
- The result is an **ionic bond**, which holds **oppositely charged ions** together.
- This explains how metal X and non-metal Y combine to form a compound XY.

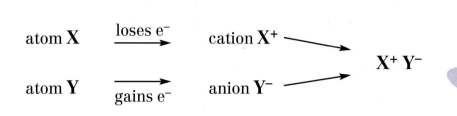

X and Y have reacted to form an **ionic** compound.

Each ion **attracts** ions of opposite charge from all directions, giving a 'giant structure' of ions.

This is what happens when **sodium reacts with chlorine**.
The sodium atoms give their outer electrons to the chlorine atoms. In this way both types of atom attain a stable outer arrangement of electrons.

The **structure** of sodium chloride is shown on **page 35**.

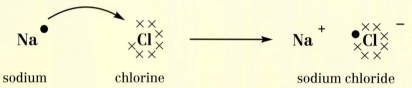

sodium   chlorine   sodium chloride

Lithium bonds to fluorine in a similar way.

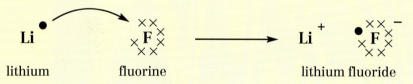

lithium   fluorine   lithium fluoride

- When magnesium reacts it has to be able to give up **two** outer electrons.

One oxygen atom can take both electrons from a magnesium atom.

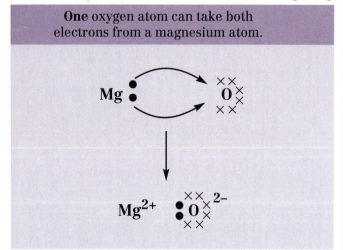

Two chlorine atoms are needed for each magnesium.

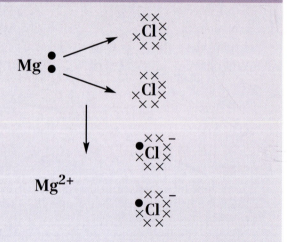

Questions

1. Reactions between which two types of atom are likely to produce ionic bonds?
2. Why do ions remain bonded together?
3. Draw 'dot–cross' (• x) diagrams to show sodium reacting with chlorine and with oxygen.
4. Which electrons are used to form bonds?
5. *Draw 'dot–cross' (• x) diagrams to show the bonds in:*

   LiCl, NaF, CaCl$_2$, AlCl$_3$, K$_2$O

Oxygen atoms have six outer electrons. When oxygen reacts each atom has to take **two** more electrons. Two sodium atoms are needed for each oxygen atom.

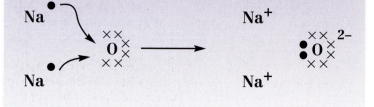

FORMULAE AND EQUATIONS I Writing formulae

The charges on ions can be used to determine the formulae of **ionic compounds**.

- In sodium chloride the single positive charge on the sodium ion (Na^+) is matched by the single negative charge on the chloride ion (Cl^-) to give the formulae NaCl.
- The formulae of ionic compounds show how many positive and negative ions are needed so that their total charges match each other.

> **One** potassium ion, K^+, needs **one** chloride ion, Cl^-, to give a formula **KCl**.
> **One** magnesium ion, Mg^{2+}, needs **two** chloride ions, Cl^-, to give a formula **MgCl$_2$**.
> **One** magnesium ion, Mg^{2+}, needs **one** oxide ion, O^{2-}, to give a formula **MgO**.
> **Two** sodium ions, Na^+, are needed for **one** oxide ion, O^{2-}, to give a formula **Na$_2$O**.

Other formulae can be worked out in the same way.

There is a table of ions on page 126.

calcium oxide	calcium chloride	aluminium chloride
(Ca^{2+} O^{2-})	(Ca^{2+} $2Cl^-$)	(Al^{3+} $3Cl^-$)
CaO	**CaCl$_2$**	**AlCl$_3$**

Some ions are made up of **groups** of **atoms**. They are sometimes referred to as **radicals** or **radical ions**. They give formulae according to the same rules as simple ions.

> **One** sodium, Na^+, needs **one** hydroxide ion, OH^-, to form sodium hydroxide, **NaOH**.
> Calcium, Ca^{2+}, needs **one** sulphate ion, SO_4^{2-}, to form calcium sulphate **CaSO$_4$**.
> Calcium, Ca^{2+}, needs **two** hydroxide ions, OH^-, to form calcium hydroxide **Ca(OH)$_2$**..
> Calcium, Ca^{2+}, needs **two** nitrate ions, NO_3^-, to form calcium nitrate **Ca(NO$_3$)$_2$**.

Brackets are used around the whole formula for a radical ion if more than one is needed.

Ca(OH)$_2$ is correct but CaOH$_2$ is incorrect.

Ca(NO$_3$)$_2$ is correct but CaNO$_3$ $_2$ is incorrect.

Iron atoms can form two types of ion. They are Fe^{2+}, iron(II) and Fe^{3+}, iron (III).

Iron(II) chloride (**Fe^{2+} $2Cl^-$**) is FeCl$_2$.

Iron (III) chloride (**Fe^{3+} $3Cl^-$**) is FeCl$_3$.

Questions
1. Use the table of ions on page 126 to work out the formulae of the compounds listed. Set out your answers in the same way as in the example given below.

Example: Give the formula for zinc oxide.   zinc oxide
(Zn^{2+} O^{2-})   **ZnO**

magnesium oxide, *potassium oxide*, magnesium chloride, *zinc sulphide*, *zinc sulphate*, sodium hydroxide, *potassium hydroxide*, *magnesium hydroxide*, *aluminium hydroxide*, *aluminium oxide*, *sodium nitrate*, *sodium sulphate*, *calcium carbonate*, *ammonium chloride*, *ammonium sulphate*.

COVALENT BONDS Covalent bonds

Covalent bonds are formed when atoms **share electrons**.

- By gaining one outer electron a hydrogen atom can complete the first electron shell.
- A chlorine atom needs one electron to make its outer shell stable.
- Hydrogen and chlorine can both attain stable electron arrangements by **sharing electrons**.

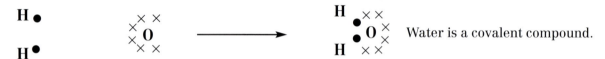

Hydrogen chloride is a **covalent** compound.

- The H and Cl atoms are held together because they **share** a pair of electrons.
- This is called a **covalent bond**.
- Non-metals can combine with other non-metals by forming covalent bonds.

Water is a covalent compound.

Carbon atoms, $_6$C 2.4, have four outer electrons. It would be difficult to lose or gain four electrons to form C^{4+} or C^{4-} ions. Carbon uses covalent bonds to form compounds.

Methane, CH$_4$, is an example of a covalent carbon compound. Four hydrogen atoms each share a pair of electrons with one carbon atom. This means that there are four covalent bonds.

- Compounds **between non-metals** are common and usually contain **covalent bonds**.
- Each **covalent** bond is formed by a **shared** pair of electrons.
- Covalent bonds can be shown in different ways.

Ammonia
NH$_3$

Carbon dioxide
CO$_2$

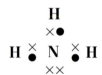

$O = C = O$

- In carbon dioxide the carbon atoms **share two pairs of electrons** with each oxygen,
- Carbon forms a **double bond to oxygen**.

A hydrogen molecule

A chlorine molecule

An oxygen molecule

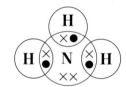

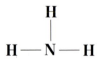

H — H

Cl — Cl

O = O

Questions

1. Name the compound formed when hydrogen combines with chlorine.
2. Why do hydrogen and chlorine react by sharing electrons?
3. What type of bond is formed when atoms share electrons?
4. How many electrons are needed to form a single covalent bond?
5. Which types of element normally combine by forming covalent bonds?
6. What is the formula of methane and why does it have four covalent bonds?
7. Draw a dot–cross diagram to show the electrons in NH$_3$.

STRUCTURE AND PROPERTIES Giant ionic structures

Ionic compounds form giant lattice structures of ions.

> Giant structures contain **very many** particles all bonded together.

- Sodium combines with chlorine by forming an ionic bond.
- The ions in sodium chloride are kept in place because the opposite charges attract.
- This leads to a crystal lattice with a **regular arrangement of ions held by strong forces**.

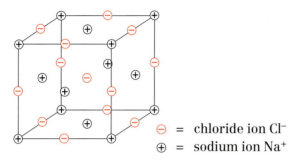

⊖ = chloride ion Cl⁻
⊕ = sodium ion Na⁺

> The structure of sodium chloride is a cubic arrangement of ions, which is why salt crystals have a cubic shape. Each sodium ion is surrounded by six chloride ions and each chloride ion is surrounded by six sodium ions. This agrees with the formula NaCl.

- The particles in a giant structure are usually bonded together in a regular pattern.
- A lot of energy is needed to break the bonds in a giant structure.
- Sodium chloride (salt) has a high melting point and a recognizable **crystal** shape.
- A **high melting point** is characteristic of all **giant structures**.

Sodium chloride, Na⁺Cl⁻, melts at 801°C.　　　Magnesium oxide, $Mg^{2+} O^{2-}$, melts at 2852°C.

- The high melting point of both these compounds indicates strong crystal lattices.
- Magnesium oxide has a higher melting point and a stronger crystal lattice.
- Magnesium and oxide ions have more charge and are smaller than sodium and chloride ions.
- Ions with a high charge and a small radius usually form stronger crystal lattices.

> Sodium chloride is a typical ionic solid.

> Sodium chloride dissoves in water.

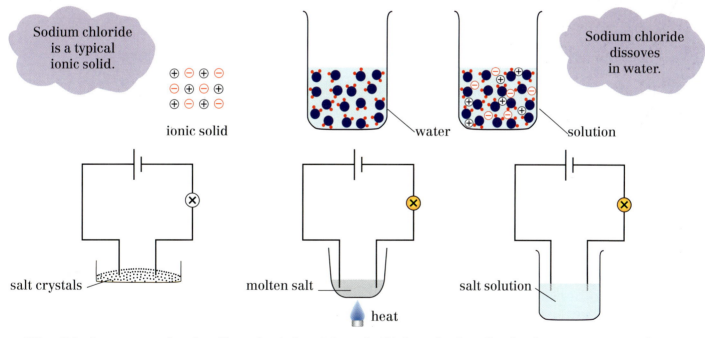

- Like all **ionic compounds** salt will **conduct** electricity **only if it is melted or dissolved**.
- Electric current is carried by the **ions** but they have to be **free to move**.
- In solids the **ions cannot move** around so sodium chloride (salt) crystals **will not conduct**.

> See pages 38 and 95

Questions

1. Why do ions stay together in ionic solids?
2. What is meant by a giant structure?
3. Why do giant structures have high melting points?
4. What is the common name for sodium chloride and what shape are its crystals?
5. Why does magnesium oxide have a stronger lattice than sodium chloride?
6. Why don't ionic solids conduct electricity?
7. What has to happen to ionic solids before they can conduct electricity and why is this necessary?
8. What carries the current when ionic substances conduct electricity?

STRUCTURE AND PROPERTIES Giant covalent structures

Giant covalent lattices have covalent bonds which extend throughout the whole structure.

- Many covalent substances occur as simple molecular structures of small molecules.
- Diamond and graphite are examples of **giant covalent structures**.
- Giant covalent structures are also referred to as **giant molecules** or **macromolecules**.

Diamond (carbon)
Each carbon atom is covalently bonded to four other carbon atoms throughout the whole structure.

The atoms are held in place by strong bonds in all directions throughout the whole structure (**hard**).

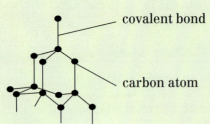

covalent bond

carbon atom

Graphite (carbon)
Each carbon atom is bonded to only three other carbon atoms and this gives a layered structure.

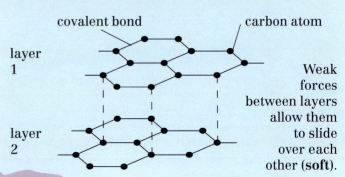

covalent bond   carbon atom

layer 1

layer 2

Weak forces between layers allow them to slide over each other (**soft**).

Diamonds and graphite are **allotropes** of carbon.

Both diamond and graphite will burn
$$C + O_2 \rightarrow CO_2$$

Diamond and graphite are **not** isotopes of carbon.

- Diamond **and** graphite are two different giant structures made by carbon atoms.
- Diamond **and** graphite can withstand temperatures of up to 3500°C without melting.
- Differences in structure give rise to some differences in properties.

DIAMOND

The hardest natural substance. Measured at 10 on the hardness scale.

A very poor conductor of electricity because all the outer electrons have formed bonds. Diamond conducts heat very well.

Density 3.5 g/cm^3

Used for cutting tools and drills because of its hardness. It also refracts light very well and does not become scratched, so is used for jewellery.

GRAPHITE

Much softer than diamond, its hardness is measured at 1–2 on the hardness scale.

One of the few non-metals to conduct electricity. Each atom only forms three bonds, leaving one free electron per carbon atom to carry current.

Density 2.3 g/cm^3

Used as electrodes and brushes in electric motors because it conducts electricity. It is used as a solid lubricant and in pencil leads because it is soft.

It is now known that carbon can form **molecules** which are like miniature versions of a soccer ball. One of these, C$_{60}$, is called **buckminster fullerene** and is made up of 20 hexagons and 12 pentagons all bonded together.

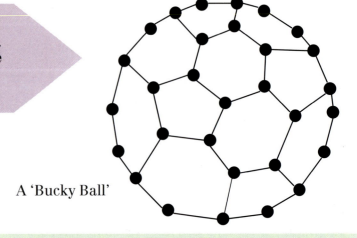

A 'Bucky Ball'

Questions
1. Which bonds occur in diamond and graphite?
2. Why are the structures of diamond and graphite unusual?
3. Graphite is soft but diamond is hard. Why?
4. Why is graphite able to conduct electricity?
5. Why do diamond and graphite have high melting points?
6. Name a giant covalent **compound**.
7. In what way is diamond **chemically** similar to graphite?

One common covalent compound with a giant structure is silica. Silica, SiO$_2$, is the main mineral in quartz. Crystals of quartz are present in sand and sandstone

STRUCTURE AND PROPERTIES   Simple molecular structures

Many covalent substances occur as small molecules and **not** giant structures.

Methane	H—C—H (with H above and below C)	CH_4	(CH₄ molecules)	methane – a gas
Water	H, H, O arrangement	H_2O	(H₂O molecules)	water – a liquid
Iodine	I—I	I_2	(I₂ molecules)	iodine – a solid
Hydrogen	H H	H_2	(H₂ molecules)	hydrogen – a gas

- In CH_4, H_2O, I_2, and H_2 the bonds do not continue beyond the individual molecules.
- Structures like this are referred to as '**molecular**' or '**simple molecular**'.
- There are usually only weak forces **between** the molecules.

All these are examples of simple molecules. Glucose molecules contain 24 atoms covalently bonded but the bonds do not extend beyond each separate molecule, therefore glucose is not a giant structure.

CO_2 H_2 Cl_2 H_2O

NH_3 N_2 HCl CH_4

C_2H_5OH $C_6H_{12}O_6$

Although each molecule is held together by covalent bonds, there are no bonds between molecules. This means that a lot of energy is **not** needed to keep the molecules apart. **Simple molecular substances tend to melt and boil at low temperatures.**

Substance	Bonding	Structure	Melting point	Boiling point	Soluble in water	Electrical conductor?
H_2O	covalent	molecular	0°C	100°C	—	poor
CH_4	covalent	molecular	–182°C	–164°C	no	no
C_2H_5OH	covalent	molecular	–117°C	79°C	yes	no
N_2	covalent	molecular	–210°C	–196°C	no	no

- In simple molecular structures there are **no ions** and **no free electrons**.
- In simple molecular structures the molecules are **not bonded** to each other.
- Simple molecular substances are **not** good conductors of electricity.

Questions

1. Give examples of covalent **compounds** which have simple molecular structures.
2. Give three examples of elements with simple molecular structures.
3. Why is methane a gas?
4. Why is the melting point of water low?
5. Why are molecular compounds not good conductors of electricity?
6. *Why is sodium chloride a solid whereas hydrogen chloride is a gas?*

STRUCTURE AND PROPERTIES Conducting electricity

An electric current involves the **movement** of charged particles.

Metals conduct in the solid state because the outer electrons are free to move.

- Current passing through a metal does **not** produce any **chemical change** in it.
- When current passes through a metal the electrons move.
- This is called **conduction**.
- Metals form giant structures using **metallic bonds**.

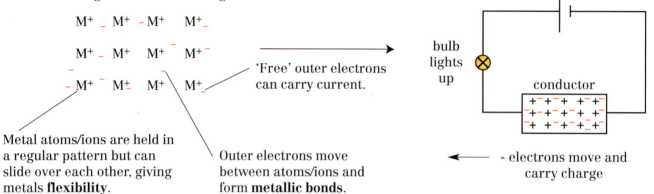

M^+ _ M^+ - M^+ M^+_

M^+ _ M^+ M^+ M^+ -

_ M^+ - M^+ M^+ M^+_

'Free' outer electrons can carry current.

Metal atoms/ions are held in a regular pattern but can slide over each other, giving metals **flexibility**.

Outer electrons move between atoms/ions and form **metallic bonds**.

bulb lights up

conductor

- electrons move and carry charge

Ionic compounds conduct electricity when melted or dissolved in water because the ions are free to move.

- Copper chloride is an ionic compound containing Cu^{2+} and Cl^- ions.
- Copper chloride will conduct electricity if dissolved in water because the ions are free to move.
- Two carbon rods called **electrodes** are dipped into the solution to complete the circuit.

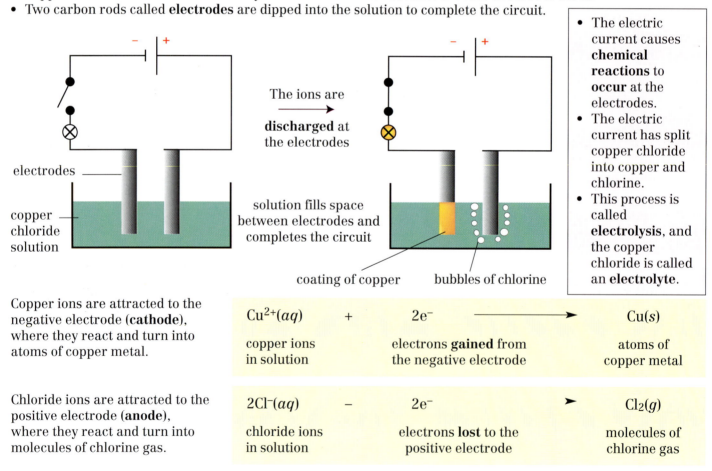

The ions are

discharged at the electrodes

electrodes

copper chloride solution

solution fills space between electrodes and completes the circuit

coating of copper   bubbles of chlorine

- The electric current causes **chemical reactions** to **occur** at the electrodes.
- The electric current has split copper chloride into copper and chlorine.
- This process is called **electrolysis**, and the copper chloride is called an **electrolyte**.

Copper ions are attracted to the negative electrode (**cathode**), where they react and turn into atoms of copper metal.

$$Cu^{2+}(aq) \quad + \quad 2e^- \quad \longrightarrow \quad Cu(s)$$

copper ions in solution

electrons **gained** from the negative electrode

atoms of copper metal

Chloride ions are attracted to the positive electrode (**anode**), where they react and turn into molecules of chlorine gas.

$$2Cl^-(aq) \quad - \quad 2e^- \quad \longrightarrow \quad Cl_2(g)$$

chloride ions in solution

electrons **lost** to the positive electrode

molecules of chlorine gas

Questions

1. What causes an electric current?
2. Which particles carry the charge in metals when they conduct?
3. Why can ionic **solids** not conduct?
4. What has to be done to ionic solids in order that they can conduct electricity?
5. Name the substances formed when copper chloride solution conducts electricity.
6. What effect does an electric current have on copper chloride solution?
7. What is meant by electrolysis?

STRUCTURE AND PROPERTIES Ions and electrolysis

Electrolysis occurs when an electric current produces a chemical reaction.

- Substances that conduct by using ions are called **electrolytes**.
- Electrolytes can be broken down into simpler substances by an **electric current**.
- The process by which an electric current breaks down a compound is called **electrolysis**.

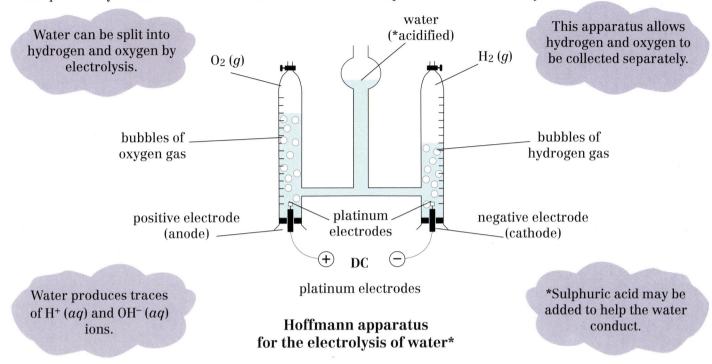

Water can be split into hydrogen and oxygen by electrolysis.

This apparatus allows hydrogen and oxygen to be collected separately.

Water produces traces of H⁺ (aq) and OH⁻ (aq) ions.

*Sulphuric acid may be added to help the water conduct.

**Hoffmann apparatus
for the electrolysis of water***

At the anode:

$$4OH^- (aq) - 4e^- \longrightarrow O_2 (g) + 2H_2O (l)$$

hydroxide ions — lose electrons → oxygen gas / water

At the cathode:

$$2H^+(aq) + 2e^- \longrightarrow H_2 (g)$$

hydrogen ions + gain electrons → hydrogen gas

- Hydrogen and oxygen may be formed when other aqueous solutions are electrolysed.
- Sodium chloride solution, for example, gives hydrogen instead of sodium (see page 91).
- This is because the hydrogen (H⁺ (aq)) ions and hydroxide (OH⁻ (aq)) may react.

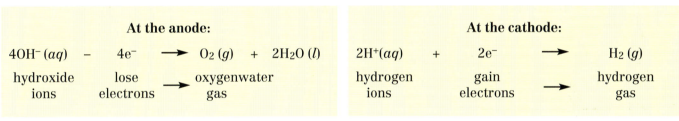

- Electrolysis of aqueous solutions shows certain patterns.
- Solutions of chlorides give chlorine at the anode.
- A metal or hydrogen is formed at the cathode.
- Oxygen is sometimes formed at the anode.

Electrolyte	At cathode	At anode	Left in solution
sodium chloride solution	hydrogen	chlorine	sodium hydroxide
copper sulphate solution	copper	oxygen	sulphuric acid
copper chloride solution	copper	chlorine	water
dilute hydrochloric acid	hydrogen	chlorine	water
dilute sulphuric acid	hydrogen	oxygen	sulphuric acid

Questions
1. Name the two gases produced by the electrolysis of water.
2. What does electrolysis do to the water?
3. What names are given to the positive and negative electrodes?
4. Why is sulphuric acid added to water before the electrolysis?
5. *Why are the two volumes of gases produced not equal?*
6. *Why is pure water not a good conductor?*
7. *Water is not a good conductor but why does it conduct at all?*
8. Name the three substances formed when sodium chloride solution is electrolysed.

REVIEW QUESTIONS Classifying materials II

1. For each of the following elements:
   (a) give the number of each type of particle
   (b) give the electron arrangement
   (c) draw a diagram to show the structure of an atom.

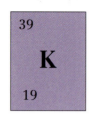

27
Al
13

39
K
19

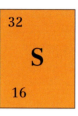

32
S
16

2. (a) What is special about the outer electron arrangements of the Noble Gases and how does it affect their reactivity?
   (b) In what way do the elements in group 1 react to change the arrangement of their outer electrons?
   (c) In what way do the elements in group 7 react to change the arrangement of their outer electrons?
   (d) When sodium reacts with chlorine what happens to the outer electron around the sodium atom, which new particles are formed, and what type of bond is formed?
   (e) Using dot–cross diagrams show how bonds form between potassium and chlorine, magnesium and oxygen, magnesium and chlorine.
   (f) Which types of atom are bonded together by ionic bonds?

3. (a) Use dot–cross diagrams to show the formation of a covalent bond between hydrogen and chlorine.
   (b) Why do hydrogen and chlorine form a covalent bond rather than an ionic bond?
   (c) Which types of atom are bonded together by covalent bonds?
   (d) What is the difference between a single bond and a double bond?
   (e) Draw a dot–cross diagram to show the electrons in a molecule of ammonia and a molecule which has a double bond.

4. (a) Why is the structure of sodium chloride described as a giant ionic structure?
   (b) Why does sodium chloride conduct electricity when melted but not when solid?
   (c) Other than by melting, how else can sodium chloride be made to conduct electricity?
   (d) What property do all giant structures have in common?

5. (a) What are the two giant structures of carbon?
   (b) Describe and explain the differences in hardness and conductivity between the two giant structures of carbon.
   (c) How is it possible to show that both these structures are made of carbon?
   (d) Name one covalent compound which has a giant structure.

6. (a) In what ways are the properties of simple molecular substances different from those with giant structures?
   (b) Give three examples of compounds with simple molecular structures.
   (c) What type of bonding usually gives rise to simple molecular structures?
   (d) Why are simple molecular substances not usually good conductors?

7. (a) Draw a diagram to show how a solution of copper chloride could be electrolysed using carbon electrodes. Label the diagram clearly to show the circuit, the anode and cathode, and the electrolyte.
   (b) Describe and explain what would be observed at each electrode.
   (c) Write ionic equations to represent the reactions at each electrode.

8. (a) Name the gases produced by the electrolysis of sodium chloride solution and explain where they come from and what happens to the sodium ions.
   (b) How is it possible to obtain sodium and chlorine from sodium chloride?

9. Use the data sheet (p.126) to write formulae for: sodium chloride, calcium chloride, sodium hydroxide, calcium carbonate, sodium carbonate, calcium hydroxide, magnesium sulphate, and sodium sulphate.

CHANGING MATERIALS CHANGES Physical changes

Physical changes do not involve chemical reactions.

- Adding salt to water does **not** change either the salt or the water into a new substance.
- Water may be frozen or boiled, but these changes do not produce a new compound.
- These are **physical** changes, and do not produce new chemical substances.

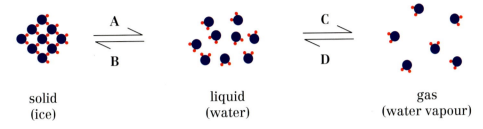

solid   liquid   gas
(ice)   (water)   (water vapour)

> Physical changes are easier to reverse than chemical changes.

- In changes A, B, C, and D the molecules of water remain unchanged.
- In A, melting, heat gives the molecules enough energy to move around.
- In C, boiling, even more energy allows the molecules break free from each other.
- In these changes the molecules are **not** split into hydrogen and oxygen.

> Separating mixtures usually involves physical changes.

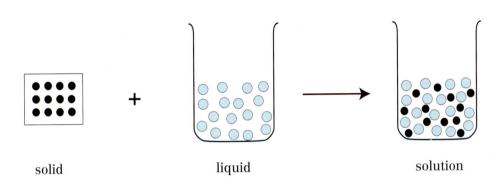

solid   +   liquid   →   solution

- In the solution the particles of solid have spread out but are not bonded to the liquid.
- There is no chemical reaction between the solid and the liquid.
- In solutions like this the **dissolved solid** is called the **solute** and the **liquid** is the **solvent**.

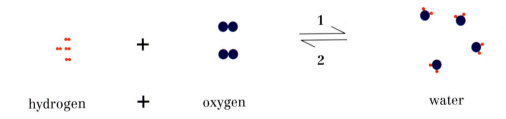

hydrogen   +   oxygen   water

- **Changes 1 and 2 are chemical changes**.
- In 1 the elements have combined to make a new compound.
- In 2 the compound has been split up to give two elements.

- Colour, melting point, solubility, and density are examples of **physical properties**.
- Being able to burn in air or react with an acid are examples of **chemical properties**.

Questions

1. Why are melting and freezing called physical changes?
2. Is boiling a physical or chemical change? Give a reason for your answer.
3. Hydrogen burns in oxygen to make water. Why is this a chemical change?
4. What happens when salt is added to water?
5. Which type of change, physical or chemical, is easier to reverse?
6. Water is a colourless liquid which boils at 100°C. Are these physical or chemical properties?
7. *What is the difference between a physical and a chemical property?*
8. *Which type of change is rusting?*

CHANGES   Chemical changes

KS3

Chemical changes take place when substances are involved in a chemical reaction. During chemical reactions **new substances** are formed.

- Adding salt to water does **not** change either the salt or the water into a new substance.
- Water may be frozen or boiled but these changes do not produce a new compound.
- These are **physical** changes and do not produce new **chemical** substances.

- If copper powder and sulphur are mixed and heated they react by combining.

| copper | + | sulphur | $\longrightarrow$ | copper sulphide |
| Cu | + | S | $\longrightarrow$ | CuS |

- If mercury oxide is heated it splits into two elements.

| mercury oxide | $\longrightarrow$ | mercury | + | oxygen |
| HgO   HgO | $\longrightarrow$ | Hg   Hg | | O_2 |

- If calcium carbonate is heated strongly it changes into two other compounds.

| calcium carbonate | $\longrightarrow$ | calcium oxide | + | carbon dioxide |
| $CaCO_3$ | $\longrightarrow$ | CaO | + | CO_2 |

| **Elements combine to form compounds.** | **Compounds split up into elements.** | **Compounds change and become different compounds.** |

- These are all examples of **chemical changes**.
- Chemical reactions produce substances which are **different** from those present at the start.
- The substances present at the start of a chemical reaction are called the **reactants**.
- The substances formed by a chemical reaction are called the **products**.

Types of reaction? See page 101.

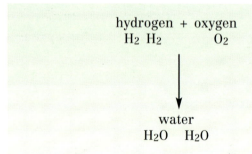

Hydrogen combines with oxygen to make hydrogen oxide. This is the reaction when **hydrogen burns** to form **water**. The new compound, water, has been made from atoms that were already present at the beginning of the reaction.

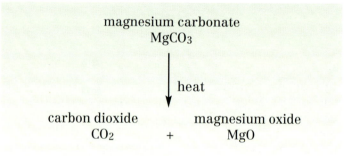

This equation shows that the two new compounds formed are made from atoms already present at the start. New atoms are not formed by chemical reactions but reactions can change the way in which they are bonded to other atoms.

Questions

1. What happens when copper and sulphur are heated together?
2. What happens to mercury oxide when it is heated?
3. Name the two new compounds formed when calcium carbonate is heated.
4. *If water splits up, which two elements are formed?*
5. *Why is water not formed when calcium carbonate is heated?*

6. Name the products formed when magnesium carbonate is heated.
7. (a) $CuS + H_2 \longrightarrow CuO + S$
   (b) $CuS + O_2 \longrightarrow Cu + SO_2$
   Which of these two reactions is not possible and why not?

OIL AND HYDROCARBONS Fossil fuels

Fossil fuels are burnt to produce energy.

- Heat, light, and transport are among the many things that require a source of energy.
- Chemical reactions which involve **burning fossil fuels** provide us with most of this energy.
- Oil, coal, and gas are major sources of fuel throughout the world.

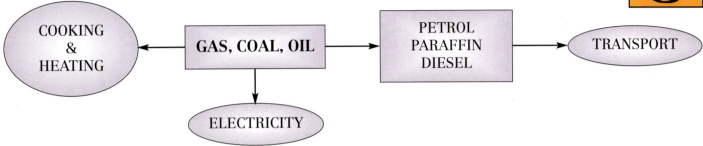

When carbon burns it forms carbon dioxide.

$$C + O_2 \longrightarrow CO_2$$

When hydrogen burns it forms water.

$$2H_2 + O_2 \longrightarrow 2H_2O$$

When compounds of hydrogen and carbon burn they form carbon dioxide and water.

e.g. methane
$$CH_4 + 2O_2 \longrightarrow CO_2 + 2H_2O$$

Reactions like this are also referred to as **combustion reactions** and the substances formed are called the products of combustion.

Hydrocarbons

- Natural gas, which is used for cooking and central heating, is methane, CH_4.
- Methane is a hydrocarbon.
- Hydrocarbons are compounds of hydrogen and carbon.
- When hydrocarbons are burnt to produce energy they also form carbon dioxide and water.

Fossil fuels

- Coal formed from the trees of ancient swamp forests.
- Coal contains a lot of carbon.
- Oil and gas formed from the remains of small organisms which lived in oceans.
- Coal, oil, and gas are found in layers of rock which are millions of years old.
- Coal, oil, and gas are fossil fuels.

Experiment   The products of combustion

The products of combustion of a **hydrocarbon** can be shown by using a **candle**, because the **wax** is made of **hydrocarbons**.

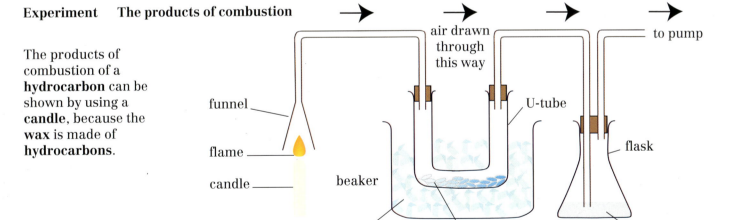

$$\text{candle wax} \xrightarrow{\text{burning}} H_2O + CO_2$$

The anyhydrous copper sulphate changes colour from white to blue as water condenses in the U-tube.

The lime water goes cloudy because it reacts with carbon dioxide.

Questions

1. What are hydrocarbons?
2. Give an example of a hydrocarbon.
3. What substances are formed when hydrocarbons burn?
4. What is needed for hydrocarbons to burn?
5. From what was coal formed?
6. *What is a combustion reaction?*
7. What is the test for carbon dioxide?
8. What is the test for water?
9. *How long did it take for coal to form?*
10. *Why are oil and gas called fossil fuels?*

OIL AND HYDROCARBONS Problems of fossil fuels

Fossil fuels are finite and non-renewable, and their combustion causes pollution.

- When hydrocarbons burn completely they form only carbon dioxide and water.
- Carbon dioxide and water vapour in the air are not directly harmful.
- When hydrocarbons burn completely they give out plenty of heat energy.
- Problems are caused by the **large scale**, world wide use of hydrocarbon fuels.

Burning hydrocarbons. **See page 49.**

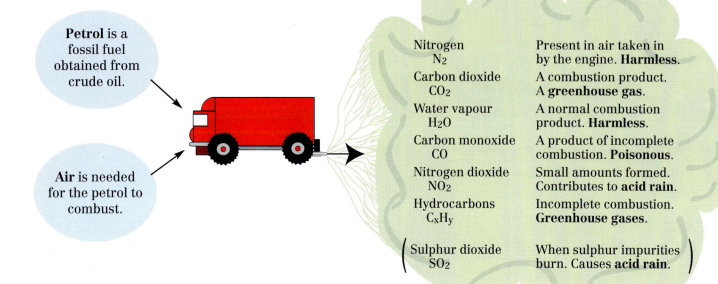

Petrol is a fossil fuel obtained from crude oil.

Air is needed for the petrol to combust.

Nitrogen N_2	Present in air taken in by the engine. **Harmless**.
Carbon dioxide CO_2	A combustion product. A **greenhouse gas**.
Water vapour H_2O	A normal combustion product. **Harmless**.
Carbon monoxide CO	A product of incomplete combustion. **Poisonous**.
Nitrogen dioxide NO_2	Small amounts formed. Contributes to **acid rain**.
Hydrocarbons C_xH_y	Incomplete combustion. **Greenhouse gases**.
Sulphur dioxide SO_2	When sulphur impurities burn. Causes **acid rain**.

Too much carbon dioxide (one possible outcome)
- Usual amount of carbon dioxide in the air is about 0.03%.
- Burning vast quantities of hydrocarbon fuels increases levels of carbon dioxide.
- Carbon dioxide traps heat from the sun.
- Global warming means that Earth's average temperature increases and ice caps melt.
- Higher sea levels cause worldwide floods.

Incomplete combustion
- Hydrocarbon fuels need a good supply of oxygen to burn completely.
- Without enough air/oxygen combustion is incomplete.
- Carbon monoxide (CO) may form instead of carbon dioxide (CO_2).
- Carbon monoxide gas is colourless, odourless, and poisonous.

Gas fires with poor ventilation

Exhaust fumes

Greenhouse Effect. (glass 'traps' heat)

glass

Short wavelength radiation from sun can get through glass.

Longer wavelength radiation emitted from warm earth cannot escape through glass.

Radiation is absorbed and soil becomes warmer.

Carbon dioxide in the air causes a greenhouse effect.

Finite resources
- Deposits of coal, oil, and gas took millions of years to form.
- Coal, oil, and gas are extracted and used in vast quantities.
- Fossil fuels are being used 1,000,000 times faster than they formed.
- Fossil fuels will run out and cannot be replaced.

Although coal is mostly carbon, impurities of sulphur compounds may be present. If these burn sulphur dioxide is formed. This can cause acid rain.

Questions
1. What is the normal percentage of carbon dioxide in the air and why might it increase?
2. Why is carbon dioxide called a greenhouse gas and what effect does this have?
3. Which poisonous gas might be produced by gas fires with poor ventilation?
4. What do the terms finite and non-renewable mean?
5. Name two gases which cause acid rain.

OIL AND HYDROCARBONS Fractional distillation

Fractional distillation is a physical process which can be used to **separate** mixtures of liquids which have different boiling points.

- Oil, petrol, paraffin, and diesel **do not mix** with water but tend to float on top.
- Liquids which do not mix are described as **immiscible**.
- Separating immiscible liquids is fairly easy.

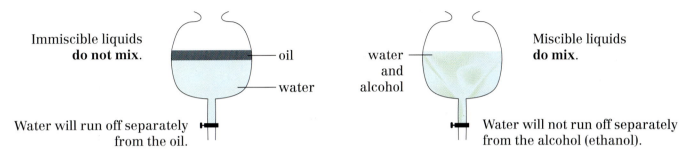

Immiscible liquids **do not mix**.

oil

water

Water will run off separately from the oil.

Miscible liquids **do mix**.

water and alcohol

Water will not run off separately from the alcohol (ethanol).

- Alcohol (ethanol) is a liquid which **does mix** with water.
- Liquids which mix completely are described as **miscible**.
- Miscible liquids may be separated if they have **different boiling points**.

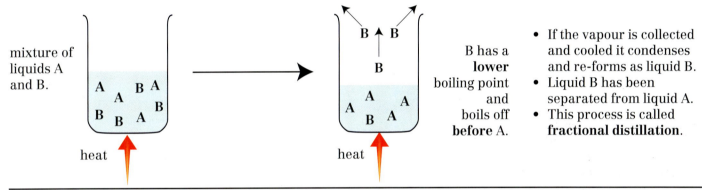

mixture of liquids A and B.

A A B A
B B A B

heat

B ▲ B

B

A A A
A B A

heat

B has a **lower** boiling point and boils off **before** A.

- If the vapour is collected and cooled it condenses and re-forms as liquid B.
- Liquid B has been separated from liquid A.
- This process is called **fractional distillation**.

Although normal distillation might work, a **fractioning column** allows better separation of the two liquids. A column is essential if more than two liquids have been mixed.

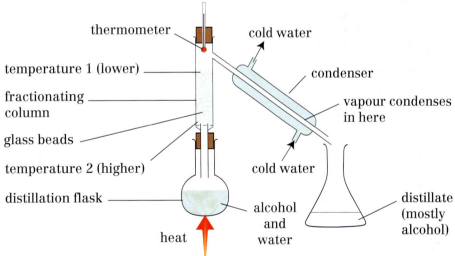

thermometer

cold water

temperature 1 (lower)

fractionating column

glass beads

temperature 2 (higher)

distillation flask

condenser

vapour condenses in here

cold water

alcohol and water

distillate (mostly alcohol)

heat

Crude oil is a mixture of liquids which have different boiling points.
Fractional distillation allows different hydrocarbons to be separated from crude oil.

Fractions from crude oil.
See page 125.

gases

gasoline
kerosene
diesel
oils
fuel oil

hot crude oil

bitumen

Fractionating column

Questions
1. What happens to oil which spills at sea?
2. Name a liquid which mixes with petrol.
3. Name a liquid which mixes with water.
4. If a mixture of two liquids is heated, why might the liquids separate?
5. What happens to a vapour if it is cooled?

OIL AND HYDROCARBONS Crude oil

Crude oil is the raw material from which fuels such as petrol and diesel are obtained.

- Crude oil (petroleum) is an oily, smelly, black-brown liquid.
- Crude oil is not a pure substance but a **mixture** of substances (mostly hydrocarbons).
- Oil companies take in crude oil and separate the different liquids **by fractional distillation**.

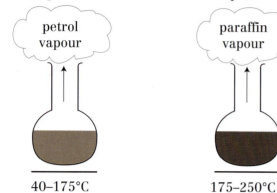

petrol vapour

paraffin vapour

diesel vapour

The liquid left becomes oilier, stickier (more viscous), thicker, and darker.

40–175°C   175–250°C   250–330°C

- This works because the different liquids in crude oil have **different boiling points**.
- The 'lighter' **fractions** boil off first at lower temperatures.
- The vapours may be collected and cooled to condense as separate liquids.

Fractionating Column. **See page 45.**

CRUDE OIL — Fractional distillation →
- Petrol
- Paraffin
- Diesel
- Others

This separation can be done as one process.

The liquids separated from crude oil are called **fractions**.

- The liquids in crude oil are mostly types of **hydrocarbon** called **alkanes**.
- They have different boiling points because their molecules have different sizes.

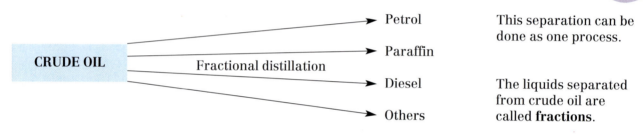

small molecules with low boiling points   →   long chain molecules with high boiling points

Pentane, C$_5$ Decane, C$_{10}$ C$_{50+}$

$CH_3CH_2CH_2CH_2CH_3$ $CH_3CH_2CH_2CH_2CH_2CH_2CH_2CH_2CH_2CH_3$ $CH_3CH_2CH_2CH_2$ ----

The distillation process does not fully separate each individual component but sorts them into **fractions** which contain molecules which are similar in size.

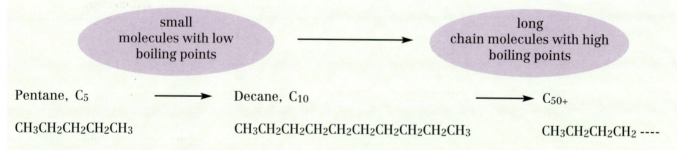

Lightest	→		FRACTIONS		→		Heaviest
Refinery gases	Gasoline (Petrol)	Kerosene (Paraffin)	Diesel	Lubricating oils	Fuel oils	Waxes/grease	Bitumen
C$_1$–C$_4$	C$_5$–C$_{10}$	C$_{10}$–C$_{15}$	C$_{15}$–C$_{20}$	C$_{20}$–C$_{30}$	C$_{30}$–C$_{40}$	C$_{40}$–C$_{50}$	C$_{50+}$

Questions

1. What does crude oil look like?
2. Is crude oil a mixture or a compound?
3. What type of substances are present in crude oil?
4. When crude oil is heated why does petrol boil off before paraffin?
5. What are fractions of crude oil?
6. What is different about the molecules in the lightest fraction compared with heavier fractions?
7. *The term octane applies to petrol. What does this suggest about petrol?*

OIL AND HYDROCARBONS Alkanes

Alkanes are **hydrocarbons** with a general formula C_nH_{2n+2}.

- Hydrocarbons are compounds of **hydrogen (H)** and **carbon (C)** only.
- Each carbon atom forms four bonds and each hydrogen atom forms one bond.
- This gives rise to a simple formula, CH_4, which is that of methane.

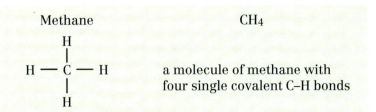

Methane CH_4

a molecule of methane with four single covalent C–H bonds

Methane is the hydrocarbon with the simplest formula. It is a colourless, flammable gas. It is the gas which is supplied to homes for cooking and heating.

- Carbon atoms can form bonds to **other** carbon atoms, thus forming **chains** of carbon atoms.
- Chains of carbon atoms give rise to many different hydrocarbons.

Methane CH_4

Octane C_8H_{18}

Methane has only one C atom per molecule.

Each octane molecule has a chain of eight C atoms.

- Methane is the first member of a series of hydrocarbons called the alkanes.
- The alkanes all have similar properties.
- As the carbon chain gets longer the molecules become larger.
- As the carbon chain gets longer the boiling points become higher.

Alkane	Formula	Structure	BP, °C	
Methane	CH_4		−164	mains gas – used for heating and cooking
Ethane	C_2H_6		−89	
Propane	C_3H_8		−42	liquefied by pressure as calor gas
Butane	C_4H_{10}		0	liquefied by pressure as lighter fuel
Pentane	C_5H_{12}		36	liquid below 36°C

Questions

1. What are hydrocarbons?
2. What is the formula of methane?
3. Describe methane.
4. For what is methane mostly used?
5. *Butane is used as lighter fuel. What causes it to remain liquid in the lighter?*
6. What name is given to the series of hydrocarbons like methane?
7. *Which fuel is associated with octane?*
8. Name a hydrocarbon which is a liquid.
9. Name two hydrocarbons which are gases.
10. *Suggest a formula for the tenth alkane.*

OIL AND HYDROCARBONS Alkenes and cracking

Alkenes are hydrocarbons which have a a general formula C_nH_{2n} **and a double bond.**

Ethene C_2H_4

- Ethene is the first member of the series of hydrocarbons called **alkenes**.
- Ethene has a **double bond** between the carbon atoms.
- Each carbon atom forms four bonds but **two** of these bonds are to the other **carbon atom**.

Propene C_3H_6

- The general formula for the alkenes is C_nH_{2n}.
- The alkene with four carbons ($n = 4$) is butene.
- **Butene** has a formula C_4H_8 and its structure includes one double bond.

Butene C_4H_8 Butane C_4H_{10}

- **Alkenes** have one **double bond**.
- **Alkenes** are **unsaturated** hydrocarbons.

- **Alkanes** have only **single bonds**.
- **Alkanes** are **saturated** hydrocarbons.

Cracking is a process in which hydrocarbon molecules with long chains are broken down into smaller molecules. Cracking is an example of a **thermal decomposition** reaction.

- Long-chain hydrocarbons are heated without air in the presence of a catalyst.
- This process is called **cracking**. The carbon chain breaks to give smaller molecules.
- Heavy fractions from crude oil are **cracked** to produce more useful smaller molecules.
- Smaller alkanes can be sold as fuels and alkenes can be used to make plastics.

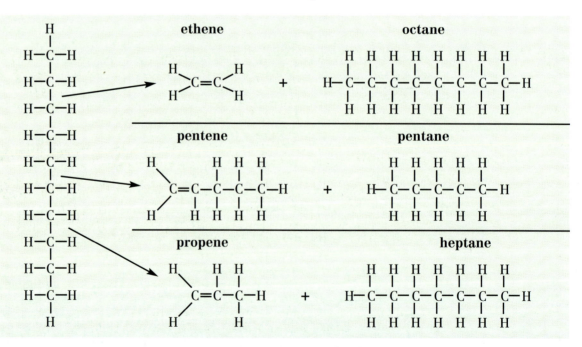

Cracking decane $C_{10}H_{22}$

The chain may crack at any position but the products will always contain an **alkane** and an **alkene**.

Questions

1. What is meant by cracking?
2. Which two types of molecule are formed when a hydrocarbon is cracked?
3. Name two products of cracking decane.
4. Give the formulae for ethene and ethane.
5. What is the difference in types of bond in alkenes compared with alkanes?
6. *Draw structures of propene and propane.*

OIL AND HYDROCARBONS   Reactions of alkanes and alkenes

Alkanes **and** alkenes burn. Alkenes also undergo **addition** reactions.

- All hydrocarbons will burn in air but some burn more cleanly than others.
- Complete combustion of a hydrocarbon produces carbon dioxide and water.

methane	+	oxygen	→	carbon dioxide	+	water
CH_4	+	$2O_2$		CO_2	+	$2H_2O$

propane	+	oxygen	→	carbon dioxide	+	water
C_3H_8	+	$5O_2$		$3CO_2$	+	$4H_2O$

- A poor supply of air can result in **incomplete combustion**, giving rise to carbon monoxide.
- Accidental deaths from carbon monoxide poisoning occur every year.

- Gas fires and gas boilers should have a good supply of air and well ventilated flues.
- It is now possible to buy carbon monoxide detectors for use in the home.

- Alkenes also burn but they give a more smoky flame than the corresponding alkanes.
- The smoke is caused by particles of unburnt **carbon** as a result **of incomplete combustion**.

ethene	+	oxygen	→	carbon dioxide	+	water
C_2H_4	+	$3O_2$		$2CO_2$	+	$2H_2O$

Ethene will burn completely if there is a plentiful supply of oxygen.

- Bromine water will distinguish alkenes from alkanes.
- Alkenes decolourise bromine water but alkanes do not.
- Bromine water changes colour from orange to colourless.
- This reaction involves the double bond present in alkenes.

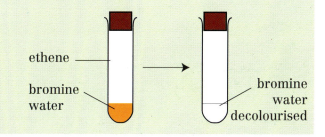

- The double bond in alkenes allows them to undergo **addition reactions**.
- Addition reactions involve other molecules **adding** across the **double bond**.

$CH_2=CH_2$	+	H–H	→	CH_3–CH_3
ethene		hydrogen		ethane

When hydrogen adds to an alkene, an alkane is formed. An unsaturated hydrocarbon becomes saturated. Margarine is made from vegetables oils by adding hydrogen to make the oils less unsaturated.

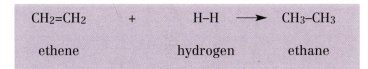

Steam can add across a double bond:				
$CH_2=CH_2$	+	H_2O	→	CH_3–CH_2OH
ethene	+	steam	→	ethanol

Halogens can add across a double bond:				
$CH_2=CH_2$	+	Br–Br	→	CH_2Br–CH_2Br
ethene	+	bromine	→	1,2-dibromoethane

Questions
1. Name the two substances produced when hydrocarbons burn completely.
2. Write a word and symbol equation for the combustion of an alkane.
3. Under what conditions might carbon monoxide be formed and why is this dangerous?
4. If ethene and ethane were each tested with bromine water, what would be observed?
5. What is formed when hydrogen adds to ethene?
6. *How could propene be converted to propane?*
7. What reacts with ethene to make ethanol?
8. *Why does ethane not undergo addition reactions?*
9. *Explain the terms 'saturated' and 'unsaturated' as applied to hydrocarbons.*

OIL AND HYDROCARBONS Polymers and polymerisation

In **polymerisation** reactions small molecules join together to make very long chains.

- Under certain conditions alkene molecules will join together to make **very long chains**.
- The compounds formed when many alkene molecules join together are called **polymers**.
- Polythene, also called poly(ethene), is an example of a polymer formed from an alkene.

Ethene (monomer)

This represents three of **many** ethene molecules.

Poly(ethene) (polymer)

This represents a small section of the polymer chain.

The molecules of ethene **add** together by using one of the bonds in the double bond.
Polymers made in this way from alkene molecules are known as **addition polymers**.

The formation of polymers – polymerisation – may be shown in different ways.

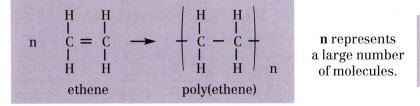

ethene   poly(ethene)

n represents
a large number
of molecules.

- Other polymers include poly(propene), poly(styrene) and **poly(vinyl chloride)** (PVC).

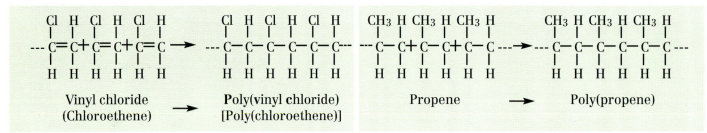

Vinyl chloride
(Chloroethene)

Poly(vinyl chloride)
[Poly(chloroethene)]

Propene

Poly(propene)

- Poly(ethene) is a soft, flexible plastic used for making plastic bags.
- Plastic bottles which contain liquids such as fabric softener, bleach or even milk are made from a high density form of poly(ethene) and many have HDPE stamped on the base.
- Other alkenes form polymers (plastics) similar to poly(ethene), but the particular properties depend on the alkene used and the conditions under which polymerisation took place.

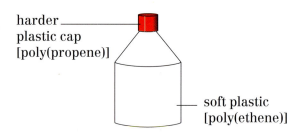

harder plastic cap [poly(propene)]

soft plastic [poly(ethene)]

plastic bottle (e.g. sterilising fluid)

- There are problems associated with the disposal of polymers.
- Polymers like this do not decay easily. They are **non-biodegradable**.
- Polymers like this do not burn easily. In some cases combustion produces toxic fumes.

Questions
1. What are polymers?
2. Why is polythene an addition polymer?
3. What environmental problems arise from the disposal or burning of plastics?
4. Name two polymers other than poly(ethene).
5. Show how ethene molecules polymerise.
6. *Why are thermoplastics not used for saucepan handles?*

- Plastics may be classified according to how they respond to heat.
- **Thermosets** are rigid plastics such as those used for light fittings or saucepan handles.
- **Thermosets** do **not** soften when they get hot.
- **Thermoplastics** are often used for packaging or plastic containers.
- **Thermoplastics** soften when heated or warmed.

METAL ORES AND ROCKS Using rocks

Rocks are substances that make up the Earth's crust.

- Granite, chalk, and basalt are all rocks.
- Rocks are usually **mixtures** of chemical substances called **minerals**.

Granite is a mixture of the minerals quartz, feldspar, and mica.

Chalk contains the mineral calcite.

Basalt often includes the minerals olivine and pyroxene.

- Most minerals are chemical compounds, many of which contain **silicon** and **oxygen**.
- Some minerals occur as **elements** but these are less common.

Mineral	Rock	Chemical description
Calcite	Chalk, limestone, marble	Calcium carbonate, $CaCO_3$
Mica (biotite)	Granite	Complex silicates, $K(Mg,Fe)_3AlSi_3O_{10}(OH,F)_2$
Feldspar (plagioclase)	Granite	Silicates, $NaAlSi_3O_8–CaAl_2Si_2O_8$
Quartz	Quartz, granite, sandstone	Silica, SiO_2

- Many **rocks are used** directly as building materials.

LIMESTONE
Is used for building walls and as an aggregate for road building.

SLATE
Is used as tiles for houses because it splits into flat sheets and is impervious to water.

GRANITE
Is a hard rock which can be cut into slabs or blocks for building. It can be polished for a decorative finish.

SANDSTONE
Can be cut into large blocks for building. Sandstone resists chemical weathering.

- Some **rocks are used** to make other substances.

LIMESTONE
Used in the extraction of iron from iron ore.
Mixed with clay and roasted to make cement.

Used to make glass.
Heated in limekilns to make quicklime.

- Some **rocks are used** for the minerals they contain.

IRON ORE provides minerals from which iron can be extracted.
- **Ores** are rocks containing minerals from which materials such as **metals** are **extracted**.

Ore	Haematite	Bauxite	Chalcopyrite	Galena	Argentite
Mineral	Fe_2O_3	Al_2O_3	$CuFeS_2$	PbS	Ag_2S
Metal	Iron	Aluminium	Copper	Lead	Silver

Questions
1. Which rock is a mixture of the minerals quartz, mica, and feldspar?
2. Which rocks are different forms of calcium carbonate?
3. *Of which elements is calcium carbonate a compound?*
4. What is slate used for and why?
5. *Granite can be used for kerbstones. Why is sandstone not suitable?*
6. What are ores?
7. Name two ores and state which metals are extracted from them.

METAL ORES AND ROCKS   Reduction and methods of extraction

Metals are extracted from their ores by a process of **reduction**.
This involves reducing the metal from its combined state back to the uncombined element.

- The figures below show the composition of the Earth's crust.
- Rocks do not contain these elements but **compounds** of these elements.
- A particular rock will be used as an ore only if the cost of extracting a metal makes it viable.

ELEMENT	oxygen	silicon	aluminium	iron	calcium	sodium	potassium	magnesium	rest
ABUNDANCE, %	46.6	27.7	8.1	5.0	3.6	2.8	2.6	2.1	1.5

- Removing oxygen from a metal oxide is one method of producing a metal.
- Losing oxygen is an example of **reduction**.
- Some metal oxides can be reduced by **heating with carbon**.

Reduction?
See page 103.

lead oxide   +   carbon $\longrightarrow$ carbon dioxide   +   lead

$2PbO(s)$ + $C(s)$ $\longrightarrow$ $CO_2(g)$ + $2Pb(s)$

Lead oxide **loses oxygen**, so is **reduced**.

Metal	Metal oxide	Reduced by carbon?
magnesium	MgO	no
aluminium	Al_2O_3	no
iron	Fe_2O_3	yes
lead	PbO	yes
copper	CuO	yes

Reduction using carbon does not work with all metal oxides. It depends on the reactivity of the particular metal. Very reactive metals do not let go of the oxygen easily whereas some, such as gold, occur naturally in an uncombined state.

- Reactive metals may be extracted by **electrolysis of the molten ore**.
- In electrolysis the reduction involves **gaining electrons**.
- Electrolysis is needed to extract sodium from sodium chloride.

Electrolysis.
See page 38.

Na (l)
molten
sodium

Sodium ions **gain electrons** from the negative electrode (cathode).

$Na^+(l)$ $Cl^-(l)$
molten
sodium chloride

The positive electrode (anode) **takes electrons** from chloride ions.

$Cl_2 (g)$
chlorine
gas

An electric current is passed through molten sodium chloride.

$$Na^+(l) \; + \; e^- \longrightarrow Na(l)$$

- Methods of extraction depend on the type of ore and the reactivity of the metal.

Metal	Reactivity	Ore	Reduction process (extraction)
sodium	high	NaCl, rock salt	electrolysis of molten ore
aluminium	high	Al_2O_3, bauxite	electrolysis of molten ore
iron	moderate	Fe_2O_3, haematite	heat with carbon in a furnace
copper	low	$CuFeS_2$, copper pyrites	roast in air
gold	very low	Au, native metal	reduction not necessary

Questions

1. Write an equation for the reaction between lead oxide and carbon.
2. Why can the reaction between lead oxide and carbon be classified as a reduction?
3. Why is it not possible to reduce aluminium oxide by heating with carbon?
4. How is sodium produced from sodium chloride and why is this method necessary?
5. *Why is reduction not necessary for the extraction of gold?*

METAL ORES AND ROCKS Extracting iron

Iron can be extracted from iron oxide ore by heating it with carbon in a Blast Furnace.

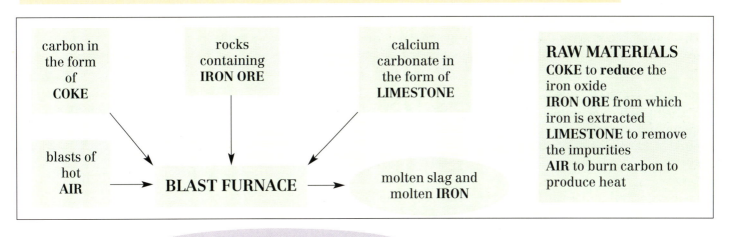

carbon in the form of **COKE**

rocks containing **IRON ORE**

calcium carbonate in the form of **LIMESTONE**

RAW MATERIALS
COKE to **reduce** the iron oxide
IRON ORE from which iron is extracted
LIMESTONE to remove the impurities
AIR to burn carbon to produce heat

blasts of hot **AIR** → **BLAST FURNACE** → molten slag and molten **IRON**

THE BLAST FURNACE

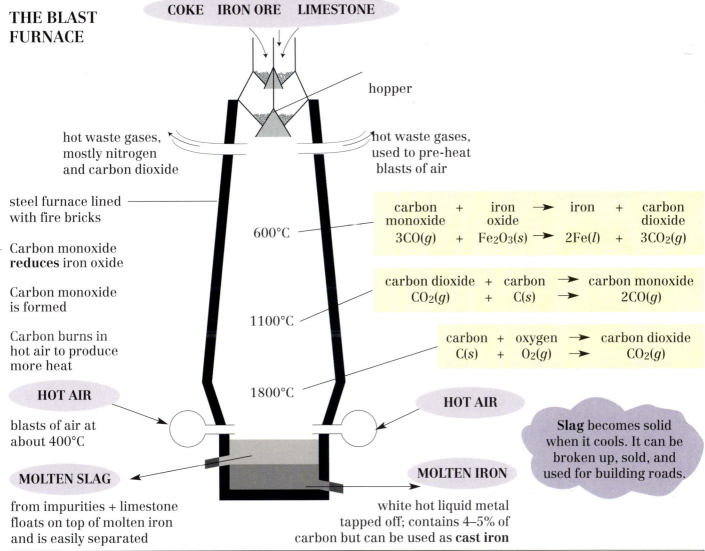

COKE   IRON ORE   LIMESTONE

hopper

hot waste gases, mostly nitrogen and carbon dioxide

hot waste gases, used to pre-heat blasts of air

steel furnace lined with fire bricks

Carbon monoxide **reduces** iron oxide

Carbon monoxide is formed

Carbon burns in hot air to produce more heat

600°C

1100°C

1800°C

carbon monoxide + iron oxide → iron + carbon dioxide
$3CO(g)$ + $Fe_2O_3(s)$ → $2Fe(l)$ + $3CO_2(g)$

carbon dioxide + carbon → carbon monoxide
$CO_2(g)$ + $C(s)$ → $2CO(g)$

carbon + oxygen → carbon dioxide
$C(s)$ + $O_2(g)$ → $CO_2(g)$

HOT AIR

blasts of air at about 400°C

HOT AIR

Slag becomes solid when it cools. It can be broken up, sold, and used for building roads.

MOLTEN SLAG

from impurities + limestone floats on top of molten iron and is easily separated

MOLTEN IRON

white hot liquid metal tapped off; contains 4–5% of carbon but can be used as **cast iron**

SLAG provides an easy method of removing silica, SiO_2, which is one of the main impurities in the iron ore. Slag is calcium silicate. It forms when CaO (a basic oxide) reacts with SiO_2 (an acidic oxide). Molten slag can be run off the surface of the molten iron.

calcium carbonate → calcium oxide + carbon dioxide
$CaCO_3(s)$ → $CaO(s)$ + $CO_2(g)$

calcium oxide + silicon oxide → calcium silicate
$CaO(s)$ + $SiO_2(s)$ → $CaSiO_3(l)$

Questions
1. Name the raw materials used to make iron.
2. Write an equation for the reduction stage.
3. Why is limestone added to the furnace?
4. *How did the blast furnace get its name?*

METAL ORES AND ROCKS Extracting aluminium

Aluminium is extracted from its ore by electrolysis.

- Aluminium does not occur naturally as the uncombined metal.
- Mostly aluminium is combined with silicon and oxygen as complex silicates in clays.
- Clays containing aluminium are **not** suitable as aluminium ores.

The main ore of aluminium is bauxite, which contains aluminium oxide and is only found in certain parts of the world.

Heating bauxite with carbon does not produce aluminium because it is too reactive and will not release the oxygen.

Aluminium is expensive because its ore is not common, and aluminium is a reactive metal making it difficult to extract.

BAUXITE ORE EXTRACTED

▼

ORE PURIFIED . . . Al$_2$O$_3$

▼

ORE MIXED WITH CRYOLITE – A COMPOUND OF SODIUM, ALUMINIUM, AND FLUORINE – AND MELTED BY ELCTRIC CURRENT

▼

LARGE ELECTRIC CURRENT PASSED THROUGH MOLTEN ORE

▼

MOLTEN ALUMINIUM METAL FORMED AT NEGATIVE ELECTRODE

The production of aluminium requires large amounts of electricity, which can be provided by hydroelectric power stations built for the extraction process.

It may be necessary to transport the aluminium ore over long distances to the extraction plant.

This will also add to the price of aluminium and probably have environmental effects on regions where the ore is obtained and the aluminium extracted.

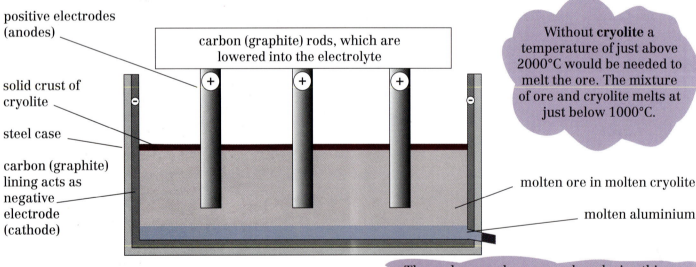

positive electrodes (anodes)

carbon (graphite) rods, which are lowered into the electrolyte

solid crust of cryolite

steel case

carbon (graphite) lining acts as negative electrode (cathode)

Without **cryolite** a temperature of just above 2000°C would be needed to melt the ore. The mixture of ore and cryolite melts at just below 1000°C.

molten ore in molten cryolite

molten aluminium

The extraction of aluminium

The carbon anodes are used up during this process because they burn in the oxygen produced.

Molten aluminium forms at the **negative electrode (cathode)** as the Al^{3+} ions gain electrons and turn into Al atoms.

$$Al^{3+}(l) + 3e^- \longrightarrow Al(l)$$

At the **positive electrode (anode)**, oxide (O^{2-}) ions lose electrons and form molecules of oxygen gas (O$_2$).

$$2O^{2-}(l) - 4e^- \longrightarrow O_2(g)$$

Questions
1. Why is aluminium not found as uncombined metal?
2. What is the main ore of aluminium?
3. Why is aluminium expensive?
4. What has to be done to aluminium oxide before it can be electrolysed?
5. What is formed at each electrode?
6. Why are the carbon anodes used up?
7. Write a word and symbol equation for the formation of aluminium atoms when aluminium oxide is electrolysed.

METAL ORES AND ROCKS Uses of iron and aluminium

Iron, in the form of steel, is the most widely used metal for engineering and structural work.

- Molten iron from the furnace can be used directly as cast iron or converted into steel.
- Steel is produced in a 'converter furnace' by adding more limestone and blowing oxygen through the molten mixture.
- Oxygen removes impurities such as sulphur, carbon, and phosphorus by changing them into gaseous oxides.
- Silicon impurities produce more slag, which is skimmed off the molten metal.

Steel	Fe,%	C,%	Properties	Uses
Wrought iron	99	0.25	malleable	chains, gates, horseshoes
Mild steel	99.5	0.5	strong, flexible	girders, rails, car bodies
Hard steel	99	1	strong, hard	cutting tools, drills, files
Cast iron	96	4	very hard, brittle	manhole/drain covers, stoves

The % values for iron and carbon are approximate.   More carbon makes the steel harder but more brittle.

- Steels with special properties are produced by controlling the amount of carbon and adding certain proportions of other metals to the molten iron.

Tungsten	Tungsten steel	A hard steel with a high melting point. Used for high-speed cutting and drilling tools
Chromium and nickel	Stainless steel	A steel which does not rust. Used for cutlery and kitchenware
Chromium	Ball bearing steel	A hard steel which resists wear
Cobalt	Magnet steel	Cobalt and iron can both be magnetised

Aluminium has a low density, is a good conductor and resists corrosion.

High-voltage overhead cables

copper

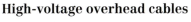
aluminium

steel

Copper is the better conductor, but would be heavier.

Aluminium core wound in steel for strength

Aircraft

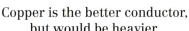

Aluminium alloy
(a steel plane would be far too heavy to fly)

Cooking foil

Shiny cooking foil reflects heat back into food.
The aluminium foil does not melt or burn.

Cans of drink

FIZZY DRINK

Alumimium does not corrode and is lightweight
so is easier to transport.

Saucepans

thermoset **plastic** handle

Aluminium conducts heat well, is not too heavy,
and does not react with hot water.

Questions
1. What is the main element in steel and what effect does carbon have on steel?
2. *What is special about stainless steel?*
3. Describe three uses of aluminium and explain why aluminium is suitable for the purpose.

METAL ORES AND ROCKS Extracting and using copper

Electrolysis is not needed to extract copper but it is used to purify copper.

- Copper is not a reactive metal.
- Chemical reactions which reduce copper compounds do not require a lot of energy.
- Once copper ore has been mined it has to be concentrated, then reduced to copper metal, which in turn has to be purified so that it conducts electricity more efficiently.

Copper ore is roasted in a limited supply of air.

copper sulphide + oxygen ⟶ copper + sulphur dioxide

$$CuS(s) \quad + \quad O_2\,(g) \quad \longrightarrow \quad Cu(l) \quad + \quad SO_2(g)$$

This is a simplified equation for the roasting and reduction.

- The impure copper solidifies in moulds as blocks of impure **blister copper**.
- Electrolysis is carried out in a solution of acidified copper sulphate as the electrolyte.
- A block of impure copper is used as the anode and a sheet of pure copper as the cathode.
- Pure copper forms on the cathode and copper ions go into solution at the anode.

Copper ore is not very common and contains only a small % of copper.

The processes of extraction and purification are complex.

Copper is required in a very pure state for electrical purposes.

Copper is a relatively expensive metal.

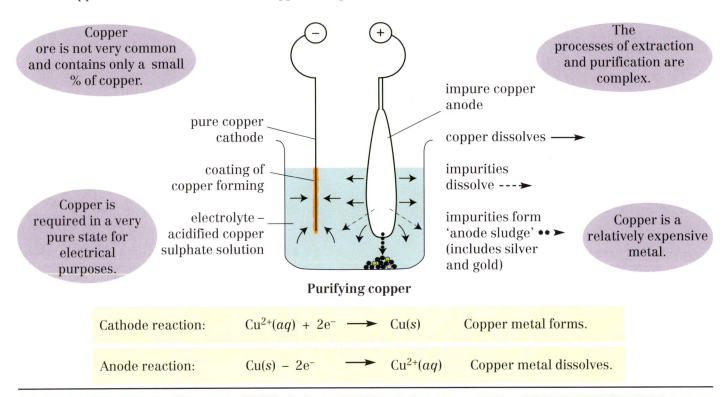

pure copper cathode

impure copper anode

coating of copper forming

copper dissolves ⟶

electrolyte – acidified copper sulphate solution

impurities dissolve ---➤

impurities form 'anode sludge' ●●➤ (includes silver and gold)

Purifying copper

Cathode reaction: $Cu^{2+}(aq) + 2e^- \longrightarrow Cu(s)$ Copper metal forms.

Anode reaction: $Cu(s) - 2e^- \longrightarrow Cu^{2+}(aq)$ Copper metal dissolves.

Copper is very ductile and is a good conductor of electricity and heat.

Copper is not a very reactive metal.

copper wire

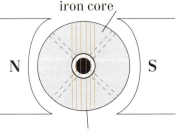

PVC insulation

iron core

coil in an electric motor many turns of thin copper wire

Brass is an alloy of copper and zinc. It is harder than copper and is used to make ornaments.

Questions
1. Explain how copper is obtained from copper sulphide.
2. Why is it necessary to use electrolysis in the purification of copper?
3. What happens at each electrode during the purification of copper?
4. What happens to the impurities which were in the copper anodes?
5. *Give two reasons why electric wires are usually made from copper.*
6. *Give two reasons why the price of copper metal is high.*

Copper and gold are coloured metals.

REVIEW QUESTIONS Changing materials I

1. Melting, dissolving, burning, rusting, evaporation, respiration.
   (a) From the above list identify two chemical and two physical changes.
   (b) Explain what is different about chemical and physical changes using one example of each type from the above list.

2. For each of the following mixtures describe a means of separating the components, and explain why each method works.
   (a) Iron from a mixture of iron and sulphur.
   (b) Salt crystals from salt solution.
   (c) Water from salt solution.
   (d) Alcohol from a mixture of water and alcohol.

3. Explain why a filter will separate a mixture of sand and water, but will not separate salt from water.

4. (a) Write an equation for the result of burning carbon in oxygen.
   (b) Explain why (a) is a chemical change.
   (c) What happens to water when it is heated at 100°C?
   (d) Why is change (c) not a chemical change?

5. (a) Name three fossil fuels and give brief details about how one of them formed.
   (b) Name the main compound in natural gas and give its formula.
   (c) Name the two products formed when hydrocarbons burn completely.
   (d) Explain how incomplete combustion might arise and why it is dangerous.
   (e) Describe two of the main concerns about the worldwide use of fossil fuels.

6. (a) What is crude oil?
   (b) Explain how crude oil is separated into fractions and why this process works.
   (c) Name three fractions from crude oil and give their uses.

7. (a) Write names, formulae, and structures for three alkanes and two alkenes.
   (b) Why are alkanes and alkenes described as hydrocarbons, what is the difference in bonding between them, and which are described as unsaturated?
   (c) How does bromine water distinguish between alkanes and alkenes?

8. Decane reacts when heated without air: decane $C_{10}H_{22}$ $\longrightarrow$ octane C_8H_{18} +
   (a) Name this process, and give the name and formula of the missing product.
   (b) Why is this process carried out on certain fractions of crude oil?

9. Ethene will undergo a reaction as shown.

   (a) Explain what has happened in this reaction and name the substance formed.
   (b) Name one other alkene which will react similarly and name the polymer formed.
   (c) Give uses for any three named polymers.
   (d) Explain why disposal of items made of polymers is a problem.

10. (a) Name the ores of aluminium and iron and give their chemical formulae.
    (b) What is the main type of reaction involved in obtaining metals from their ores?
    (c) Explain why some metals are extracted by heating with carbon and some require electrolysis, and state, with a reason, which method is likely to be more expensive.

11. (a) Name the three raw materials, other than iron ore, used in the blast furnace.
    (b) Write an equation for the reaction in which iron ore produces iron.
    (c) Why is slag also produced in the blast furnace and how is it separated from the iron?
    (d) What is the main use of iron?

12. Aluminium is extracted from its ore by electrolysis.
    (a) What is the ore mixed with to help it melt?
    (b) Write an ionic equation for the formation of aluminium at the cathode.
    (c) State what is formed at the anode, and explain why the anodes are used up during the extraction and how this affects the design of the electrolysis cell.
    (d) Give two uses for aluminium and the properties which are important for these uses.

13. (a) Explain why it is necessary to purify copper after it has been extracted.
    (b) Draw and label a simple diagram to describe how copper is purifed.
    (c) Write ionic equations for the reactions at each electrode when copper is purified.

THE ATMOSPHERE Current composition of the atmosphere

The composition of the atmosphere has been almost unchanged for the past 200 million years.
Approximately 1/5 of dry air is oxygen and 4/5 nitrogen.

Experiment

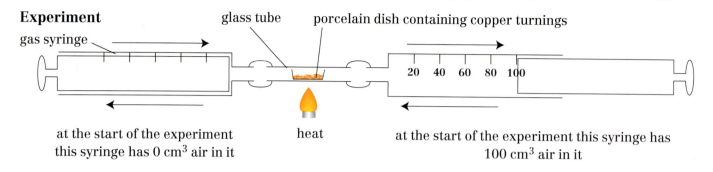

glass tube   porcelain dish containing copper turnings

gas syringe

at the start of the experiment
this syringe has 0 cm³ air in it

heat

at the start of the experiment this syringe has
100 cm³ air in it

As the copper turnings are heated, air is pushed from one syringe to the other so that it flows over the copper.

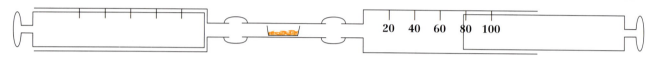

The volume of the air becomes **less** as the copper **reacts** with the **oxygen** in the air.

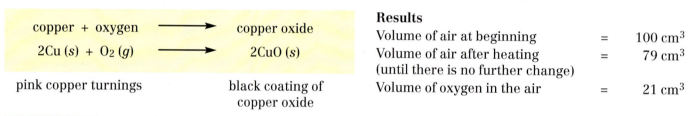

copper + oxygen $\longrightarrow$ copper oxide

$2Cu\,(s) + O_2\,(g) \longrightarrow 2CuO\,(s)$

pink copper turnings   black coating of
copper oxide

Results

Volume of air at beginning	=	100 cm³
Volume of air after heating (until there is no further change)	=	79 cm³
Volume of oxygen in the air	=	21 cm³

Many processes affect the composition of the air, e.g. respiration, photosynthesis, burning of fossil fuels.
To maintain the composition of the atmosphere there must be a balance between the processes which add and
remove gases – **gases** are **recycled**.

Approximate composition of the air

Gas	% by volume
nitrogen	78
oxygen	21
argon	1
carbon dioxide	0.03
other Noble Gases e.g. helium, neon	traces
water vapour	amount varies

Uses of nitrogen
• Production of ammonia (Haber Process) and from ammonia, nitric acid, fertilizers, dyes, explosives
• Liquid nitrogen is used as a refrigerant
• Nitrogen is pumped into empty fuel tanks to reduce fire risk

Uses of oxygen
• With ethyne (acetylene) in welding torches
• In hospitals in breathing apparatus
• In the manufacture of steel

Uses of carbon dioxide
• Solid carbon dioxide is used to keep food frozen
• Fire extinguishers
• Fizzy drinks (the bubbles are carbon dioxide)

Questions

1. Which gas makes up approximately 4/5 of the Earth's atmosphere?
2. Name three processes which may affect the composition of the air.
3. Give three uses for each of the following gases:
   (a) nitrogen
   (b) oxygen
   (c) carbon dioxide.
4. For approximately how many years has the composition of the atmosphere remained constant?
5. Which substance combines with copper when it is heated in air?
6. *Name four components of the atmosphere that are elements.*
7. *Name two components of the atmosphere that are compounds.*
8. *Nitrogen gas does not allow substances to burn in it. Why is it important that air should contain nitrogen as well as oxygen?*

THE ATMOSPHERE   Evolution of the atmosphere

EARLY ATMOSPHERE

gases present in the atmosphere today

gases not present in the atmosphere today

Origin of the early atmosphere
One view is that the early atmosphere was formed from intense volcanic activity. Another theory is that it developed from a pre-volcanic primary atmosphere consisting of hydrogen and helium.

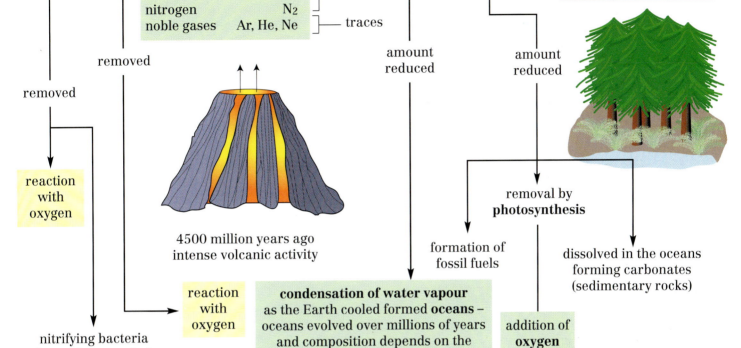

large amounts —

| carbon dioxide | CO_2 |
| water vapour | H_2O |

| ammonia | NH_3 | small |
| methane | CH_4 | amounts |

| nitrogen | N_2 | traces |
| noble gases | Ar, He, Ne | |

removed

removed

reaction with oxygen

4500 million years ago intense volcanic activity

amount reduced

amount reduced

reaction with oxygen

removal by **photosynthesis**

formation of fossil fuels

dissolved in the oceans forming carbonates (sedimentary rocks)

condensation of water vapour as the Earth cooled formed **oceans** – oceans evolved over millions of years and composition depends on the balance shown below

addition of **oxygen**

nitrifying bacteria changed **ammonia** to **nitrates**

denitrifying bacteria changed nitrates to **nitrogen**

input of dissolved salts from weathering of rocks

removal of dissolved salts by shell formation in marine animals

removal by chemical reactions to give sea floor sediments – crystallization giving salt deposits

evolution of **photosynthesizing organisms** on land and in the sea produced O_2.

increase in O_2 gave fewer habitats for anaerobic organisms

Atmosphere of some other planets
Mercury: no atmosphere
Venus: mainly carbon dioxide
Mars: carbon dioxide, argon

Ozone
The ozone layer developed by the action of ultra violet light (uv) on oxygen. Ozone filters out harmful ultra violet light from the sun. Some scientists consider that the destruction of the ozone layer by pollutants, e.g. CFCs, may lead to an increase in the risk of skin cancer.

Questions
1. What are the theories about the origin of the early atmosphere?
2. What was the importance of each of the following processes in changing the composition of the early atmosphere:
   (a) cooling of the Earth   (c) nitrifying bacteria
   (b) photosynthesis   (d) denitrifying bacteria?
3. Which gases found in the early atmosphere are not present today?
4. Which gases found in the early atmosphere are still present today?
5. Why can the oceans be described as 'reservoirs' for carbon dioxide?
6. Which gases are found in the atmospheres of Mercury, Venus, and Mars?
7. *What does 'anaerobic' mean?*
8. *Draw up a table in two columns to compare the composition of the early atmosphere with that of today.*

Oceans act as 'reservoirs' for carbon dioxide, both absorbing and releasing the gas. Limestone rocks are 'reservoirs' for lime and carbon dioxide.

There is a complex and natural balance between carbon dioxide in the air, carbon dioxide and lime dissolved in the sea, the formation of limestone and changes in climate. Limestone rocks dissolve and recrystallise easily and resist weathering in dry climates.

THE ATMOSPHERE   The carbon cycle

The carbon cycle is one of the cycles which help to maintain the composition of the atmosphere. The amount of carbon dioxide in the atmosphere depends on the rates at which carbon dioxide is added and removed from the atmosphere.

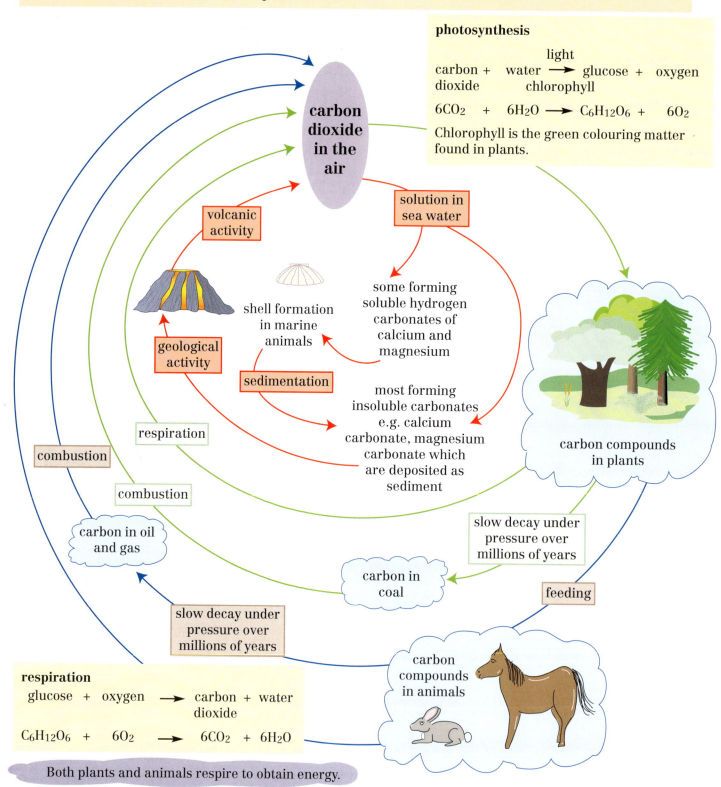

photosynthesis

$$\text{carbon dioxide} + \text{water} \xrightarrow[\text{chlorophyll}]{\text{light}} \text{glucose} + \text{oxygen}$$

$$6CO_2 + 6H_2O \longrightarrow C_6H_{12}O_6 + 6O_2$$

Chlorophyll is the green colouring matter found in plants.

carbon dioxide in the air

volcanic activity

solution in sea water

some forming soluble hydrogen carbonates of calcium and magnesium

shell formation in marine animals

geological activity

sedimentation

most forming insoluble carbonates e.g. calcium carbonate, magnesium carbonate which are deposited as sediment

carbon compounds in plants

respiration

combustion

combustion

carbon in oil and gas

slow decay under pressure over millions of years

carbon in coal

feeding

slow decay under pressure over millions of years

respiration

$$\text{glucose} + \text{oxygen} \longrightarrow \text{carbon dioxide} + \text{water}$$

$$C_6H_{12}O_6 + 6O_2 \longrightarrow 6CO_2 + 6H_2O$$

carbon compounds in animals

Both plants and animals respire to obtain energy.

Questions

1. Give two ways in which carbon dioxide is removed from the atmosphere.
2. Give three ways in which carbon dioxide is returned to the atmosphere.
3. *What is meant by 'global warming'?*
4. *List some of the ways in which human activities may increase the amount of carbon dioxide in the air.*
5. *What could be done to help cut down the amount of fossil fuels being burned?*
6. *How do you think media presentations might affect people's views on global warming?*

Global warming?
See page 44.

However, some scientists think there may be a decrease in temperature because of the **increase in particulates** (matter and dust) in the atmosphere from burning fuels, mining, and volcanic eruptions.

THE EARTH   Types of rock

Rocks may be classified as sedimentary, igneous or metamorphic.

Granite is a hard rock with a speckled appearance in which crystals of the minerals quartz, feldspar, and mica are clearly visible.

Igneous rocks
- Molten magma rises from beneath the Earth's crust, cools, and solidifies.
- Interlocking **crystals** fused together in a random arrangement are a clue to rock formed from molten material.
- There are no fossils.

Minerals are the chemical elements or compounds which are present in rocks.

Sandstone is often identified as a red-orange rock in which individual grains of quartz can be seen. The colour is caused by other minerals such as iron oxide but sandstones vary in colour and hardness depending on which minerals are present.

Limestone often contains fossils.

Sedimentary rocks
- Layers of sediment, for example, beneath oceans are buried and **compressed** into rock.
- Fragments or **grains**, rather than crystals, held together are a clue to rocks formed from sediment.
- Fossils may be present.

- Cementation contributes to the consolidation of sediment into rock.
- Water percolates through the spaces (pores) between the particles.
- Material dissolved in the water precipitates out and binds the particles together.
- Cementing agents include calcite, quartz, and iron oxide.

Gneiss is a hard, coarse-grained rock in which alternate bands of light and dark minerals are visible.

- **Metamorphic rocks**
- Igneous and sedimentary rocks are changed by **heat** and **pressure** without being melted.
- Cleavage planes and crystals arranged in layers are clues to rocks formed by heat and pressure acting on other rocks.
- There are no fossils.

- Rocks are solid mixtures of minerals which make up the Earth's crust.
- Magma is a molten mixture of minerals which forms at certain places beneath the crust.
- If magma cools it becomes solid and forms crystals of minerals (**crystallization**).
- Most rocks form inside the crust but layers of rock may be pushed to the surface (**uplift**).
- Surface rocks are worn away to fragments which may settle as sediment (**sedimentation**).
- If sediments remain buried long enough they may consolidate (harden) as new rock.
- Heat and pressure can cause minerals to recrystallise or change into new minerals.

Questions
1. Why is granite speckled?
2. What are minerals? Give two examples.
3. How are sandstones formed?
4. What is meant by sedimentary rock?
5. What are rocks and where do they form?
6. Which type of rock might have fossils?
7. How was granite formed and what name is given to all rocks formed this way?
8. How are metamorphic rocks formed?
9. Name and describe a metamorphic rock.

THE EARTH The rock cycle

The Earth's crust is a **solid** mixture of rocks which are involved in a continual **cycle** of processes during which existing rocks are changed and new rocks are formed.

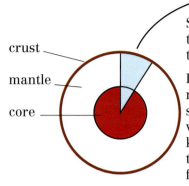

crust

mantle

core

Solid rock 6–10 km thick under oceans. Up to 65 km on continents.

Rock similar to crust ~~but~~ more dense. Although solid, the rock circulates very slowly. About 3000 km thick, with temperatures ranging from 1500 to 5000°C.

Mostly iron and nickel. The outer core is molten. The inner core is solid owing to the immense pressure it is under. The radius is about 3000 km and the temperature is about 5500°C.

The rocks on Mount Everest show evidence that they were formed beneath an ancient ocean.

continental crust

ocean

oceanic crust

mantle

mantle

lithosphere

The lithosphere is the **rigid** outer layer of crust **and** upper mantle down to a depth of 50–100 km. The lithosphere is broken into large sections called **plates**.

- Rocks are continually changing but many of the changes occur (very slowly) over millions of years.
- Rocks formed beneath the Earth's surface may be seen as cliffs, hills or mountains.
- Rocks may be worn away, and forces in and beneath the Earth's crust can push layers slowly upwards.

THE ROCK CYCLE

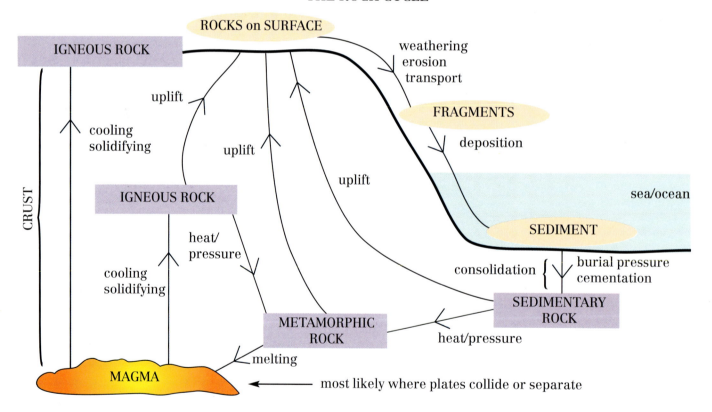

Questions

1. Name the three main layers of the Earth and state which one is made from rocks.
2. What is magma and how does it become igneous rock?
3. How are sedimentary rocks formed?
4. Most rocks form beneath the Earth's surface so why are many rocks visible?
5. What is the connection between metamorphic and the other two types of rock?
6. What happens to rocks which are pushed beneath the Earth's crust?
7. *Into which metamorphic rock can limestone be changed and what causes this change?*

THE EARTH Weathering

Weathering is a general term for a range of processes by which surface rocks are changed.

- **Physical weathering** causes rocks to be broken down into smaller fragments.
- Changes in **temperature** can cause physical weathering.

Water runs into cracks in a rock.

Water freezes and expands.

Cracks in rock widen.

Rock breaks

Weathering and erosion are different. **Weathering** occurs *in situ* but **erosion involves movement** and is a process by which rocks are gradually worn away. Grains of sand blown by the wind, and running water can erode rocks.

Cold shrinks rock

Very high daytime temperatures

Very cold at night

Heat expands rock

- **Chemical weathering** causes changes in the chemical composition of minerals in rocks.
- Water, carbon dioxide, and oxygen are common agents of chemical weathering.

The rate of chemical weathering is affected by particle size and temperature.

feldspar (in granite)

water and carbon dioxide ↓

clay

Water and carbon dioxide form carbonic acid.

calcium carbonate (in limestone)

water and carbon dioxide ↓

'dissolved' as calcium hydrogencarbonate

iron silicates

water and oxygen ↓

'rust'

Sediments are fragments of rock which have moved and settled into a new position.

Fragments may vary from large boulders to fine clay particles.

ROCKS

↓ weathering

FRAGMENTS

↓ transport erosion deposition

SEDIMENT

Fragments may be moved by wind, water, ice or gravity.

Fragments may move a few metres or thousands of kilometres.

Rock fragments which have not been carried far will be angular, as in breccia.

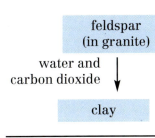

fine grains

angular fragments

Grains of sand which have been carried long distances by water will be rounded, as in most sandstones.

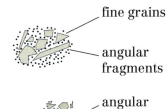

angular grains

rounded grains

Grains of sand which have been carried by the wind are spherical and highly polished, as in dune sandstone.

spherical grains

Questions
1. Which type of process changes rocks?
2. What is the result of weathering?
3. Why does freezing water split rocks?
4. What is the difference between weathering and erosion?
5. Why does rainwater dissolve limestone?
6. *Where is hot and cold weather most likely to affect rocks?*
7. *How are sand and pebbles formed?*
8. *What is meant by transport and deposition?*

THE EARTH   Evidence from sedimentary rocks

The type, size, and shape of particles in a sedimentary rock provide clues about its origins.

Particle	Size (mm)	Sediment	Rock
fine	0.004 to 0.0625	silt	shale
medium	0.0625 to 2	sand	sandstone
coarse	2 to 64	gravel/pebbles	conglomerate

- Sorting refers to the range of different sized grains present in sedimentary rock.
- A narrow range of sizes is 'well sorted' whereas many sizes is 'poorly sorted'.
- Particle size and sorting can give clues about how a sediment was transported, how far it travelled, and the conditions under which it settled out.

- **Fossils** found in sedimentary rocks may provide clues about age, climate, and conditions.
- Burial in an undisturbed environment of accumulating muddy sediment is necessary.
- Many fossils are preserved **imprints** or **patterns** of the remains of **past organisms**.
- For an animal this is likely to be the pattern of its **shell** or **skeleton**.
- The shell or skeleton might be replaced by a mineral, or the pattern imprinted on a rock.

- Certain rock layers (wherever they occur) contain identical fossils.
- Fossils in rock layers above and below are different.
- The succession of fossils matches the succession of rock layers.
- Rocks containing similar fossils are of similar age.

- Trilobites were marine invertebrates which first appeared about 500–550 million years ago and became extinct about 250 million years ago.
- During the 300 million years they existed many different species evolved and died out.
- Each species lasted, on average, 10 million years.
- The presence of a particular trilobite fossil provides evidence for the age of a rock.

Trilobites

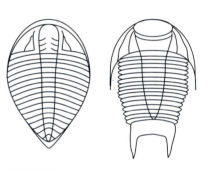

LIMESTONE	Formed from the shells and skeletons of small sea creatures, evidence of which can often be seen as fossils still present in the limestone. Some limestone rocks in Britain were once parts of coral reefs. This is evidence for movement of the Earth's crust.
CHALK	Is a white rock containing a high percentage of calcium carbonate. It does not appear to contain fossils but powerful microscopic examination shows that it is made from minute calcite discs called coccoliths originally present in algae. The extensive chalk deposits in the south-east of England formed during a period when most of Britain was covered by a warm chalk sea in which coccoliths were about the only material settling. Up to 1000 m of chalk was deposited at a rate of 30 years for each millimetre.
COAL	Formed from the vegetation of ancient swamp forests; the hardness and amount of carbon are clues about the age of the coal.

- Layers or **beds** of sediment may vary in thickness and sequences may be repeated.
- **Bedding planes**, where one layer forms on another, are usually horizontal and parallel.
- Beds laid at angles on dunes or ripples give rise to a pattern called **cross bedding**.
- Particles in a bed ranging from coarse on the bottom to fine on top show **graded bedding**.
- Graded bedding is associated with sediment which has slipped down a sloping sea bed.

Ripple marks may form along a beach, on sand dunes or on the bottom of a stream.

Symmetrical ripples indicate backwards and forwards motion of waves.

Asymmetric ripples indicate the water was flowing in one direction.

Questions

1. What are sediments?
2. *How do sediments become rock?*
3. Which parts of an organism are most likely to become fossilised?
4. Why do limestones often contain fossils?
5. *Approximately how many years did it take to form a chalk cliff 100 m high?*
6. How are sediments transported?
7. *What allows graded bedding to occur and why are the fine particles on top?*

THE EARTH   Evidence from rock layers

The sedimentary rocks seen at the present time were formed millions of years ago and the oldest usually occur in the deepest layers.

- Changes in conditions usually produced different types of sediment.
- Deposits for limestones formed beneath seas which were warm and shallow.
- Coal formed from deposits laid down in ancient swamp forests.

SEDIMENTARY LAYERS

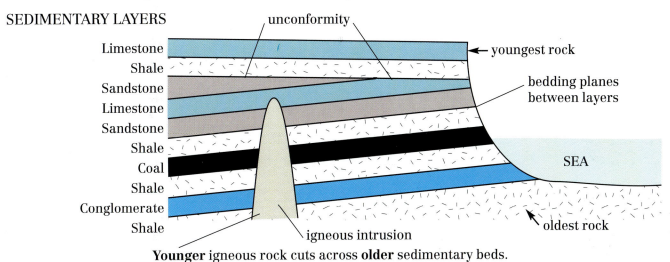

Younger igneous rock cuts across older sedimentary beds.

- Generally layers of sediment were laid down horizontally.
- There are regions where the layers are tilted or folded or have slipped past each other.
- Tilts, folds, and faults provide evidence of powerful forces in the Earth's crust.

An anticline of porous rock beneath non-porous rock may contain deposits of oil or gas trapped in the porous rock.

Upright folds

Pressure from the sides can cause layers to fold.

Folds are common in regions where continental plates collide.

Overturned folds

Tilted layers

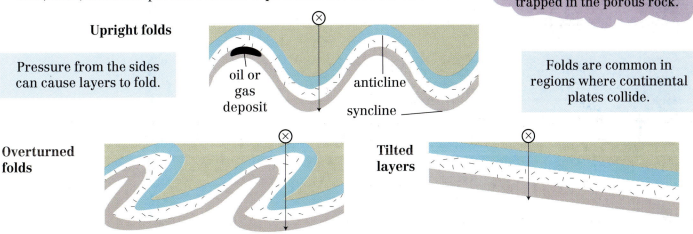

- Faults arise where rock layers have broken and slipped past each other.
- Faults arise where pressure in the crust has caused rocks to fracture rather than fold.
- Faults usually indicate that the vertical forces are greater than the horizontal forces.

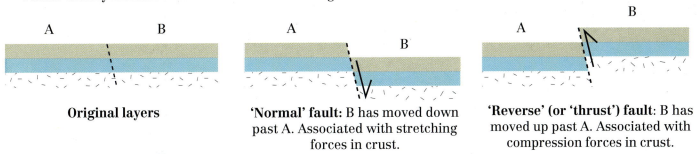

Original layers

'Normal' fault: B has moved down past A. Associated with stretching forces in crust.

'Reverse' (or 'thrust') fault: B has moved up past A. Associated with compression forces in crust.

Questions
1. Under what different conditions were deposits that became limestones and coal formed?
2. Why do faults and folds suggest there are powerful forces in the Earth's crust?
3. Where one rock cuts across another what does this suggest?
4. What are bedding planes, faults, and folds?

THE EARTH   Evidence from igneous and metamorphic rocks

Igneous rocks are formed from **molten** material.

- When magma moves **into the crust** it may cool and solidify as **intrusive igneous rock**.
- If magma reaches the **surface** it produces flows of volcanic **lava** or ash deposits.
- Lava and other material from volcanoes form **extrusive igneous rock**.

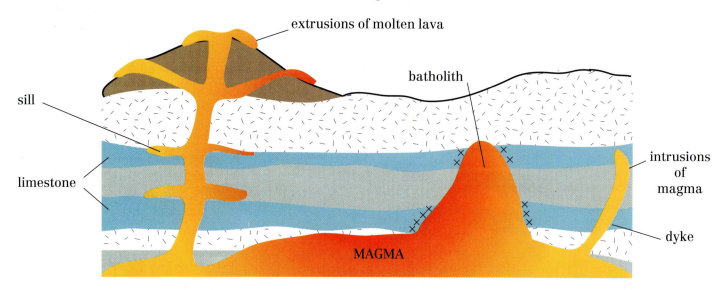

At × limestone is changed to marble by **contact metamorphism**.

- **Intrusive igneous rocks** formed **beneath** the Earth's surface.
- They contain larger crystals because cooling and crystallization were slow.
- **Granite** is an intrusive igneous rock in which the crystals are clearly visible.

- **Extrusive igneous rocks** formed **above** the Earth's surface.
- They contain small crystals because cooling and crystallization were rapid.
- **Basalt** is an extrusive igneous rock in which the small crystals are not easy to see.

Metamorphic rocks are formed by changing existing rocks **without melting**.

- **Metamorphism** is the process by which metamorphic rocks are formed.
- It involves the action of **heat** and **pressure** on sedimentary or igneous rocks.
- It may involve **recrystallization** of existing minerals or the formation of new minerals.
- It produces different changes depending on the proportions of heat and pressure.

LIMESTONE (sedimentary) ⟶ Heat and pressure ⟶ MARBLE (metamorphic)

The main mineral in both these rocks is calcium carbonate (calcite). This is taken as evidence that marble was made from limestone. This change does not involve the formation of new minerals but the calcite recrystallises. Impurities in the limestone are pushed out of the new crystals, which are whiter and larger. The result is a harder, whiter rock with veins of darker impurities which produce a 'marbling' effect in some of the rocks produced.

- **Contact metamorphism** occurs around igneous intrusions.
- Heat from the magma causes recrystallization, or new minerals to form.
- Contact metamorphism is caused mostly by heat, which 'bakes' the surrounding rock.

Many metamorphic rocks are formed by Regional Metamorphism. **See page 67.**

Questions
1. Explain the difference between intrusive and extrusive igneous rocks.
2. Why does granite have larger crystals than basalt?
3. What are sills and dykes?
4. What two factors cause metamorphism?
5. Marble is a contact metamorphic rock. Suggest how marble is formed from limestone.

THE EARTH Plate tectonics

There is a lot of **geological activity** at boundaries between moving **plates**.

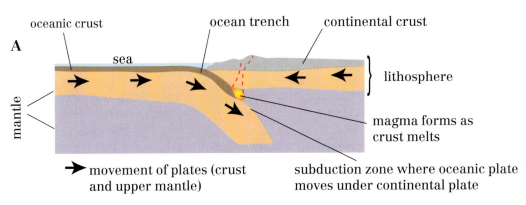

A

oceanic crust — ocean trench — continental crust — sea — lithosphere — mantle — magma forms as crust melts

➡ movement of plates (crust and upper mantle)

subduction zone where oceanic plate moves under continental plate

The lithosphere, a rigid outer layer of the Earth, consists of crust and the uppermost section of the **mantle**.
The lithosphere is not a continuous shell around the Earth but is broken into large sections called **plates**.
The plates are continually moving, very slowly, carried by **convection currents** in the mantle.

- Diagram **A** shows an oceanic plate colliding with a continental plate.
- The more dense oceanic crust is pushed beneath the lighter continental crust.
- This is a **destructive plate boundary** because crust moves down into the mantle.
- The descending plate is heated, partly by friction, and some of it melts forming magma.
- The magma moves upwards through the crust.
- Earthquakes and volcanoes are common along this type of plate boundary.

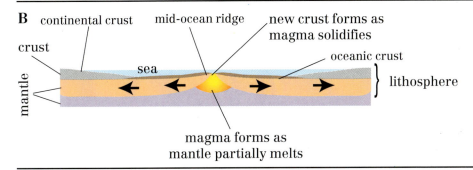

B

continental crust — mid-ocean ridge — new crust forms as magma solidifies — crust — sea — oceanic crust — lithosphere — mantle

magma forms as mantle partially melts

- Diagram **B** shows two oceanic plates are moving apart.
- The reduction in pressure allows molten magma to form.
- The magma moves upwards and solidifies as new crust.
- This is known as a **constructive plate boundary** because new crust is formed.

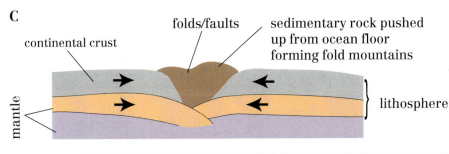

C

folds/faults — sedimentary rock pushed up from ocean floor forming fold mountains — continental crust — lithosphere — mantle

- Diagram **C** shows two continental plates colliding.
- The crusts are the same density so they push against each other, producing upward forces.
- These forces can push sedimentary rock up thousands of feet above sea level forming very large mountain ranges such as the Himalayas.

The Himalayas began to form 30 million years ago when the Indian and Eurasian plates collided. The pressure and heat in the crust resulting from these collisions was sufficient to cause metamorphism over extensive areas. This is known as **regional metamorphism**. Different degrees of heat and pressure produce different types of metamorphic rock.

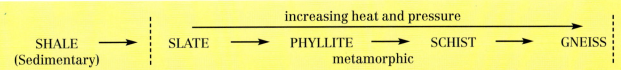

increasing heat and pressure

SHALE (Sedimentary) ⟶ SLATE ⟶ PHYLLITE ⟶ SCHIST ⟶ GNEISS

metamorphic

Slate is a **low-grade** metamorphic rock. The clay minerals in the shale have changed into **chlorite** and small colourless **mica** crystals. As the degree of metamorphism increases, different minerals are formed and the chlorite and mica develop as larger flakes, which are clearly visible in schist. Other changes include cleavage planes (obvious in slate) and banding of different minerals within the rock, both of which are a consequence of the pressure and indicate the direction of this pressure. Gneiss is a **high-grade** metamorphic rock.

Questions

1. Describe the main features of a destructive plate boundary.
2. Explain what is meant by regional metamorphism and describe how it occurs.
3. *Explain, with examples, the difference between high-grade and low-grade metamorphism.*

THE EARTH   Limestone

Limestone consists of calcium carbonate and this determines its chemical properties.

- Limestone effervesces with hydrochloric acid and is dissolved by it, producing carbon dioxide in the process:

calcium carbonate   +   hydrochloric acid $\longrightarrow$ calcium chloride   +   carbon dioxide   +   water

$$CaCO_3(s)   +   2HCl(aq)   \longrightarrow   CaCl_2(aq)   +   CO_2(g)   +   H_2O(l)$$

- Carbonic acid is formed when **carbon dioxide** in the air mixes with **rainwater**.
- Carbon dioxide is not very soluble in water.
- Carbonic acid is not a strong acid.
- Limestone reacts with carbonic acid and so is dissolved by rainwater.
- The reaction between limestone and carbonic acid is very slow.

| These reactions are reversible. | water + carbon dioxide $\rightleftharpoons$ carbonic acid $H_2O(l) + CO_2(g) \rightleftharpoons H_2CO_3(aq)$ | Carbon dioxide may be absorbed or released. |

| Limestone may dissolve or recrystallise. | calcium carbonate + carbonic acid $\rightleftharpoons$ calcium hydrogencarbonate $CaCO_3(s) + H_2CO_3(aq) \rightleftharpoons Ca(HCO_3)_2(aq)$ |

LIMESTONE   calcium carbonate $CaCO_3(s)$

↓(heat)   ↓(heat)

QUICKLIME   calcium oxide $CaO(s) + CO_2(g)$

↓(water)   ↓(+ $H_2O(l)$)

SLAKED LIME   calcium hydroxide $Ca(OH)_2(s)$

↓(filter)   ↓(filter)

LIMEWATER   calcium hydroxide solution $Ca(OH)_2(aq)$

- Limestone only undergoes thermal decomposition if it is heated strongly.
- Calcium oxide is a basic oxide which neutralises the acidity in soil.
- Calcium oxide is not too basic and does not wash off the soil quickly.
- Carbon dioxide reacts with limewater to form a white precipitate which makes limewater go cloudy ('milky').

| calcium hydroxide + carbon dioxide $\longrightarrow$ calcium carbonate + water $Ca(OH)_2(aq) + CO_2(g) \longrightarrow CaCO_3(s) + H_2O(l)$ | Carbon dioxide turns limewater cloudy by forming **insoluble calcium carbonate**. |

| calcium carbonate + carbon dioxide + water $\longrightarrow$ calcium hydrogencarbonate $CaCO_3(s) + CO_2(g) + H_2O(l) \longrightarrow Ca(HCO_3)_2(aq)$ | Excess carbon dioxide causes the limewater to go clear again as **soluble** calcium hydrogencarbonate forms. |

Questions
1. Name the substances formed when $CaCO_3$
   (a) is heated
   (b) reacts with hydrochloric acid.
2. Write an equation for the formation of carbonic acid.
3. Why does limestone react with rainwater and why is this reaction slow?
4. Give chemical names for limestone, quicklime, slaked lime, and limewater.

REVIEW QUESTIONS   Changing materials II

1. Explain the importance of each of the following processes in the carbon cycle:
   (a) photosynthesis
   (b) respiration
   (c) combustion.

2. Complete the following simplified diagram of the carbon cycle by drawing in arrows and naming the processes involved.

carbon dioxide in the air		carbon compounds in plants

carbon in sediment deep in the Earth	carbon in fossil fuels	

carbon in calcium carbonate sediment in the sea		carbon compounds in animals

3. Explain, with examples, the differences between igneous, metamorphic, and sedimentary rocks.

4. Draw a simple rock cycle to show the connections between the three types of rock.

5. Use examples to explain the difference between physical and chemical weathering.

6. (a) Draw a simple flow diagram to show how rocks form sediment.
   (b) State two ways in which fragments of rock can be transported.
   (c) What evidence is provided by the size and shape of rock fragments in sedimentary rock?

7. (a) What are fossils?
   (b) How may fossils help to determine the age of certain types of rock?
   (c) If similar rocks from different parts of the world contain the same types of fossil, what does this suggest?
   (d) Why are there often lots of fossils in limestone?

8. (a) What is a bedding plane?
   (b) What is graded bedding and under what circumstances will it have occurred?
   (c) What evidence might be obtained from ripple patterns preserved in some rocks?
   (d) What gave rise to the large chalk deposits in southern England?

9. (a) Draw simple diagrams to show what is meant by a fold and a fault.
   (b) What evidence is provided by folds or faults?

10. (a) Explain, with examples, the differences between intrusive and extrusive igneous rocks.
    (b) Explain what is meant by contact metamorphism.
    (c) If limestone undergoes contact metamorphism what is likely to form and what evidence will there be that the two types of rock are connected?

11. (a) What are the Earth's plates?
    (b) Describe what happens when an oceanic plate meets a continental plate.
    (c) Why are large mountain ranges likely to form where two continental plates collide?
    (d) What type of metamorphism occurs where continental plates collide?
    (e) What is the difference between a low-grade and a high-grade metamorphic rock?

12. (a) Write equations to show the changes from limestone to quicklime to slaked lime to limewater.
    (b) Explain why carbon dioxide turns limewater cloudy.
    (c) What happens if excess carbon dioxide passes through limewater?
    (d) What is carbonic acid?
    (e) Write an equation to show the action of rainwater on limestone.

13. In what way do oceans and limestone rock act as reservoirs for carbon dioxide and why is this important for our atmosphere?

PATTERNS OF BEHAVIOUR

METALS   Reactions with oxygen 1

When metals react with oxygen, oxides are formed. Some metals are more reactive than others. Metals can be placed in an order of reactivity based on how they react with oxygen.

Experiment   **Burning magnesium in oxygen**

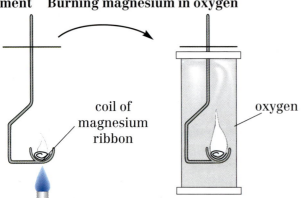

coil of magnesium ribbon

oxygen

When magnesium is heated over a Bunsen burner, it burns with a bright white flame.

Magnesium burns more brightly in oxygen than in air. A white powder (**magnesium oxide**) remains in the combustion spoon.

Experiment   **Heating copper in oxygen**

copper turnings

oxygen

Copper does not burn in air or oxygen but becomes coated in a black powder (**copper oxide**).

magnesium + oxygen $\longrightarrow$ magnesium oxide

copper + oxygen $\longrightarrow$ copper oxide

Magnesium is **more reactive** than copper. When magnesium and copper combine with oxygen they both become **oxidised**. These are **oxidation reactions**.

this symbol means **highly flammable**

SODIUM METAL

this symbol means **corrosive**

Sodium metal is stored under oil – this metal is so reactive that it would react with oxygen in the air without heating.

Summary of reactions of metals with oxygen		
Metal	Reaction	Product
sodium	burns vigorously after gentle heating	sodium oxide
magnesium	burns easily with a bright white flame to form a white powder	magnesium oxide
iron	burns in the form of a powder or wire wool, giving off sparks and forming a black solid	iron oxide
copper	does not burn, but a black powder forms on the surface of the copper	copper oxide
gold	does not react even when heated strongly	

Questions

1. Name the compound formed when magnesium burns in oxygen or air.
2. Name the compound formed when copper is heated in oxygen or air.
3. What does the term 'oxidised' mean?
4. Which metal, magnesium or copper, is more reactive?
5. Why is sodium metal stored under oil?
6. What do the hazard signs on a jar containing sodium metal mean?
7. *From the information in the table place the five metals in an order of reactivity, most to least.*
8. *Would you expect the mass of a piece of magnesium to change when it burns in air or oxygen? Explain your answer.*

METALS   Reactions with oxygen II

When metals combine with oxygen there is a change in mass.

Experiment   Heating magnesium in air

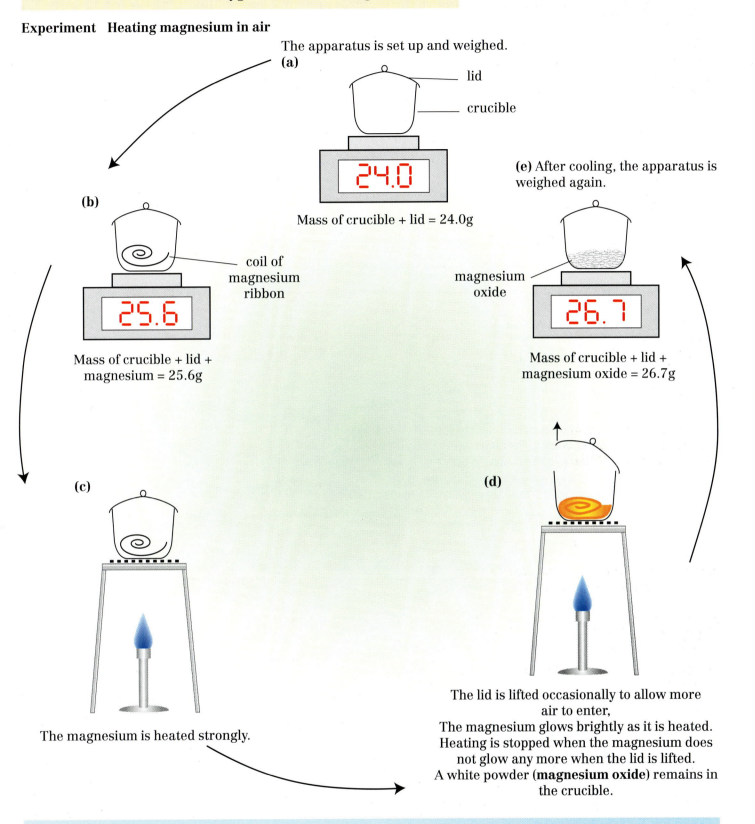

The apparatus is set up and weighed.
(a)

lid

crucible

24.0

Mass of crucible + lid = 24.0g

(e) After cooling, the apparatus is weighed again.

(b)

coil of magnesium ribbon

25.6

Mass of crucible + lid + magnesium = 25.6g

magnesium oxide

26.7

Mass of crucible + lid + magnesium oxide = 26.7g

(c)

(d)

The magnesium is heated strongly.

The lid is lifted occasionally to allow more air to enter,
The magnesium glows brightly as it is heated.
Heating is stopped when the magnesium does not glow any more when the lid is lifted.
A white powder (**magnesium oxide**) remains in the crucible.

Questions

1. What is the mass of magnesium used in the experiment?
2. What is the mass of magnesium oxide produced in the experiment?
3. What is the gain in mass?
4. How can you explain this gain?
5. Why must the crucible lid be lifted during heating?
6. Apart from the change in mass, what evidence is there that a chemical reaction has taken place?
7. *Why must the lid be lifted carefully and only slightly?*
8. *In another experiment 12.0g of magnesium combined with 8.0g of oxygen. What mass of magnesium oxide would be formed from: (a) 2.4g of magnesium (b) 4.8g of magnesium?*

METALS   Reactions with oxygen III — Rusting

Rusting is a type of corrosion (the reaction of a metal with air, water, and other substances in its surroundings).

Experiment   The conditions necessary for rusting

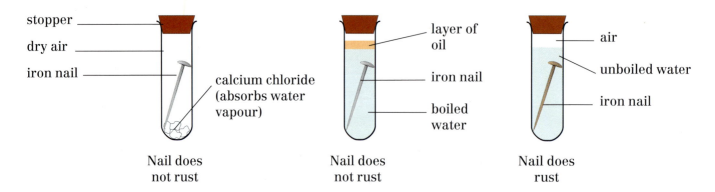

For rusting to take place, water and oxygen must be present. The iron is **oxidised**.

iron + oxygen $\longrightarrow$ iron oxide (rust)

Preventing rusting

Stainless steel is expensive and not suitable for many structures. Using iron is cheaper, but rusting must be prevented. Some methods involve coating iron so that it does not come into contact with air and water.

1. *Painting* (often with paints containing lead and zinc) e.g. iron bridges

2. *Grease and oil* e.g. machinery

3. *Plastic coating* e.g. garden furniture, dish racks

4. *Plating with an unreactive metal* e.g. chromium may be used on car bumpers, tin may be used to coat 'tin cans' made of steel

Sydney Harbour Bridge. Over a quarter of a million litres of paint were needed for the first three coats of the steel bridge, which was opened in 1932.

Other methods involve using a more reactive metal, which will become oxidised in place of the iron.

5. *Galvanizing* (coating the iron with zinc) e.g. dustbins, buckets

6. *Sacrificial protection* e.g. a bar of magnesium may be attached to the hull of a ship. It corrodes instead of the iron (steel) and must be replaced when it wears away.

Questions

1. What substances need to be present if iron is to rust?
2. In the experiment, what does calcium chloride do?
3. What is the chemical name for rust?
4. Why would stainless steel be unsuitable for building a bridge?
5. *Why does the nail not rust in the boiled water?*
6. *Why does the nail rust in unboiled water?*
7. *Give one reason why rusting should be prevented.*
8. *Suggest methods of rust prevention for each of the following and explain how each method works:*
   (a) a farm gate   (d) a refrigerator
   (b) an oil storage tank   (e) iron railings.
   (c) a door hinge

72

When metals react with acids, hydrogen is produced. Some metals are **more reactive** than others. Metals can be placed in an **order of reactivity** based on how they react with **acids**.

Experiment   Magnesium and sulphuric acid

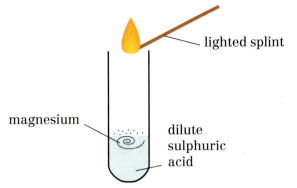

lighted splint

magnesium

dilute sulphuric acid

The gas ignites with a squeaky pop – **this is the test for hydrogen**.

The mixture fizzes and a colourless gas is given off. The magnesium disappears leaving a colourless solution.

magnesium + sulphuric acid ⟶ magnesium sulphate + hydrogen

Summary of reactions of metals with dilute acids

NOTE: In this table, the results for magnesium and zinc have been shown with both dilute hydrochloric and dilute sulphuric acid.

Metal	Acid	Observations	Products
magnesium	hydrochloric acid	fizzes rapidly, a colourless gas is given off, the magnesium disappears leaving a colourless solution, the contents of the test tube become warm	magnesium chloride + hydrogen
magnesium	sulphuric acid	fizzes rapidly, a colourless gas is given off, the magnesium disappears leaving a colourless solution, the contents of the test tube become very warm	magnesium sulphate + hydrogen
zinc	hydrochloric acid	fizzes steadily, a colourless gas is given off, the zinc dissolves slowly leaving a colourless solution	zinc chloride + hydrogen
zinc	sulphuric acid	fizzes steadily, a colourless gas is given off, the zinc dissolves slowly leaving a colourless solution	zinc sulphate + hydrogen
iron	sulphuric acid	fizzes slowly, a colourless gas is given off, the iron dissolves slowly leaving a pale green solution	iron sulphate + hydrogen
copper	sulphuric acid	no reaction	

Questions

1. Describe the chemical test for hydrogen.
2. What are the products when magnesium reacts with (a) dilute hydrochloric acid (b) dilute sulphuric acid?
3. In the reaction between magnesium and dilute sulphuric acid, what evidence is there that a chemical reaction has taken place?
4. Complete the following word equations:
    (a) magnesium + hydrochloric acid ⟶   (d) zinc + hydrochloric acid ⟶
    (b) zinc + sulphuric acid ⟶   (e) magnesium + sulphuric acid ⟶
    (c) iron + sulphuric acid ⟶
5. From the information in the table place the four metals in an order of reactivity, most to least.
6. *How does this compare with the order of reactivity obtained by looking at reactions of metals with oxygen?*
7. *How would you expect sodium to react with dilute acids? Give a reason for your answer.*

METALS   Reactions with water

When metals react with water, hydrogen is produced. Some metals are **more reactive** than others. Metals can be placed in an **order of reactivity** based on how they react with **water**.

Experiment   Calcium and water

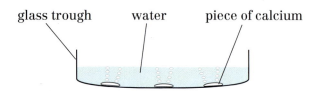

The calcium sinks. The pieces rise to the surface as bubbles of a colourless gas are given off from beneath them and then sink again as the gas is released. The calcium disappears leaving a slightly cloudy mixture.
If universal indicator solution is added to this the indicator turns violet, showing that an alkaline solution has been formed.

Experiment   Magnesium and water (steam)

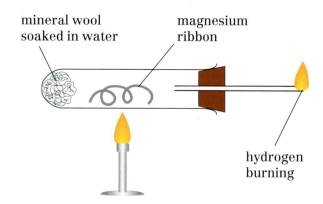

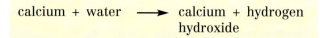

calcium + water $\longrightarrow$ calcium + hydrogen
hydroxide

magnesium + water $\longrightarrow$ magnesium + hydrogen
(as steam)   oxide

Summary of reactions of metals with water

Metal	Observations	Products
potassium	reacts violently, floats, melts, fizzes, a colourless gas is given off and ignites, moves very quickly across the surface of the water and the potassium disappears to leave an alkaline solution	potassium hydroxide + hydrogen
sodium	reacts vigorously, floats, melts, fizzes, a colourless gas is given off, moves quickly across the surface of the water and the sodium disappears to leave an alkaline solution	sodium hydroxide + hydrogen
calcium	reacts steadily, sinks, rises to the surface and then sinks again, fizzes, a colourless gas is given off, and the calcium disappears to leave a cloudy alkaline mixture	calcium hydroxide + hydrogen
magnesium	slow reaction with cold water, but reacts with steam to form a white powder	magnesium oxide + hydrogen
copper	no reaction	

Questions

1. (a) When calcium pieces are added to water, what do you see happening?
   (b) Explain these observations.
2. From the information in the table place the five metals in an order of reactivity, most to least.
3. Why do you think steam rather than water has to be used with magnesium?
4. Complete the following word equations:
   (a) potassium + water   (b) sodium + water
5. *How does the order of reactivity obtained in these reactions compare with that obtained by looking at reactions of metals with oxygen and dilute acids?*
6. *When calcium reacts with water a cloudy mixture is produced. What does this tell you about calcium hydroxide?*
7. *The summary does not include zinc.*
   (a) *Under what conditions would you expect zinc to react with water?*
   (b) *What would be the products of the reaction?*
   (c) *Where would this place zinc in the reactivity series?*
   (d) *Write a word equation for the reaction.*

METALS   Reactions with oxides
Oxidation and reduction

Metals may react with oxides of other metals – the metals **compete** for the oxygen.
A more reactive metal takes oxygen away from a less reactive metal.

Experiment   Heating iron and copper oxide

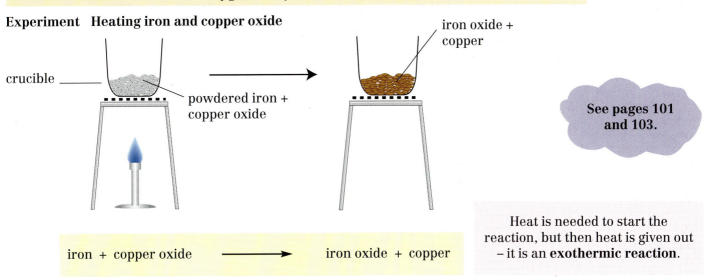

crucible — powdered iron + copper oxide

iron oxide + copper

See pages 101 and 103.

iron + copper oxide	$\longrightarrow$	iron oxide + copper

Heat is needed to start the reaction, but then heat is given out – it is an **exothermic reaction**.

Another **competition** reaction is used to repair railway lines. Powdered aluminium is mixed with iron oxide and a magnesium fuse is used to start the reaction.

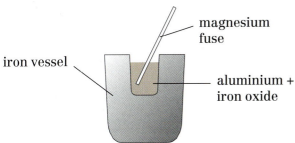

magnesium fuse

iron vessel

aluminium + iron oxide

So much heat is given out that the iron formed in the reaction is molten and can be run into gaps between the railway lines.

aluminium + iron oxide	$\longrightarrow$	aluminium oxide + iron

This is sometimes called the **Thermit Reaction**.

Oxidation and Reduction

Oxidation can be defined as the addition of oxygen.

Reduction can be defined as the removal of oxygen.

Look again at the equation for the reaction between iron and copper oxide.

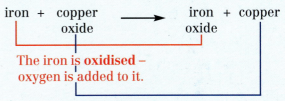

iron + copper oxide $\longrightarrow$ iron + copper oxide

The iron is **oxidised** – oxygen is added to it.

The copper is **reduced** – oxygen is removed from it.

Oxidation and **reduction** occur together. These types of reactions are called REDOX REACTIONS

Oxidising and Reducing Agents
An **oxidising agent** adds oxygen. A **reducing agent** removes oxygen.
In this reaction, copper oxide is an oxidising agent and iron is a reducing agent.

Questions

1. What are the products when powdered iron and copper oxide are heated together?
2. When iron and copper oxide react together, which chemical is:
   (a) oxidised   (c) an oxidising agent
   (b) reduced   (d) a reducing agent?
3. Which metal is more reactive – iron or copper?
4. What are the products when aluminium and iron oxide are heated together?
5. *When aluminium and iron oxide react together, which chemical is:*
   (a) oxidised   (c) an oxidising agent
   (b) reduced   (d) a reducing agent?
6. *Will the following chemicals react together when heated*
   (a) Iron and magnesium oxide
   (b) Magnesium and copper oxide
   (c) Copper and magnesium oxide?
7. *Write word equation(s) for any reaction(s) which will take place in Question 6.*
8. *Why is it true to say that oxidation and reduction occur together?*
9. *Why do you think a thick iron vessel is used for the Thermit Reaction?*

METALS Reactivity series

Using reactions with oxygen
reactions with water
reactions with dilute acids
reactions with oxides
} metals can be placed in an order of reactivity or **Reactivity Series**.

most reactive

potassium

sodium

calcium

magnesium

aluminium

zinc

iron

lead

- - - - - - - - - - - - - - -

(hydrogen)

copper

silver

gold

least reactive

The more reactive a metal, the more stable its compounds are.

Metals above this line displace **hydrogen** from **dilute acids**.

Metals below this line **do not react** with **dilute acids**.

Although hydrogen is not a metal, it is sometimes included in the **Reactivity Series**.

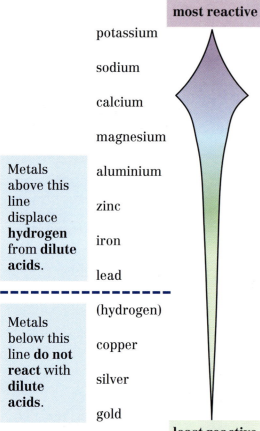

Pink

Shoes

Can

Make

A

Zebra

In

London

(**H**ave)

Clean

Socks –

Good

Learning a sentence like this one, or making up one of your own, can help you to remember the order of metals in the series.

Questions

1. Which metals in the series displace hydrogen from dilute acids?
2. (a) Name the two most reactive elements in the series.
   (b) Would you expect them to occur naturally as elements or compounds? Give a reason for your answer.
3. Which metals in the series have the most stable compounds?
4. *The following table describes some of the reactions of five metals. The metals are copper, iron, magnesium, sodium, and zinc.*
   Identify each of the metals, A–E, and then place them in order of reactivity, most to least.

Metal	Reaction with oxygen	Reaction with dilute hydrochloric acid	Reaction with water
A	does not burn, but becomes coated with a black powder	no reaction	no reaction with water or steam
B	burns with a bright white light to form a white powder	vigorous reaction	reacts slowly with cold water, but more rapidly with steam
C	burns vigorously with a yellow flame to form a white powder	violent reaction	melts and fizzes to produce a colourless gas
D	does not burn easily, but when it does so, forms a black solid	slow reaction	slow reaction with steam
E	does not burn easily, but when it does so, forms a white solid when cool	quite a slow reaction unless powdered	quite a slow reaction with steam

5. *Give word equations for each of the reactions of B and C with oxygen, dilute hydrochloric acid, and water.*
6. *Write out the Reactivity Series using chemical symbols instead of names.*

METALS Displacement reactions

A more reactive metal will displace a less reactive metal from a solution of its salts. Reactions between metals and dilute acids are another type of displacement reaction. Many metals displace hydrogen from a dilute acid.

Experiment   Iron and copper sulphate solution

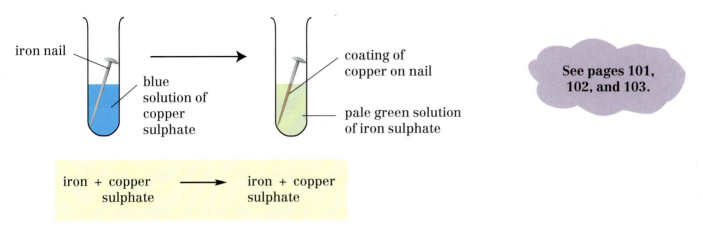

See pages 101, 102, and 103.

iron + copper sulphate ⟶ iron + copper sulphate

Iron displaces copper from copper sulphate solution. Iron is **more reactive** than copper.

Experiment   Copper and silver nitrate solution

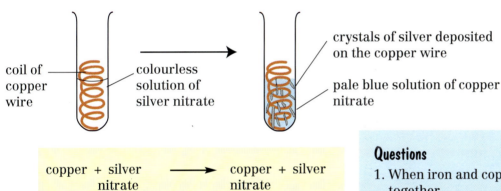

copper + silver nitrate ⟶ copper + silver nitrate

Copper displaces silver from silver nitrate solution. Copper is **more reactive** than silver.

Experiment Copper and calcium nitrate solution

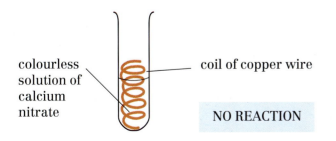

Copper does not displace calcium from calcium nitrate solution. Calcium is **more reactive** than copper.

Questions

1. When iron and copper sulphate solution react together
   (a) Which metal is displaced?
   (b) Which metal is more reactive?
2. When copper and silver nitrate react together
   (a) Which metal is displaced?
   (b) Which metal is more reactive?
3. Why is there no reaction when copper wire is placed in a solution of calcium nitrate?
4. *Will the following chemicals react together*
   (a) magnesium and copper sulphate solution
   (b) copper and magnesium sulphate solution
   (c) zinc and copper nitrate solution
   (d) lead and copper nitrate solution?
   In each case explain your answer.
   Give word equations wherever a reaction can take place.
5. *Why is hydrogen given off when magnesium and zinc are added to dilute acids?*
6. *Why is there no reaction when copper is added to a dilute acid?*
7. *Of the metals listed in the Reactivity Series only copper, silver, and gold occur naturally as elements. Suggest a reason for this.*

ACIDS, BASES AND SALTS Indicators and pH

Acidic solutions turn litmus solution red and blue litmus paper red.
Acids are corrosive when concentrated and can eat away skin, cloth, and metals.
Acidic solutions have a pH less than 7.
Acids are composed of hydrogen and other non-metals.

Alkaline solutions turn litmus solution blue and red litmus paper blue.
Alkalis are corrosive when concentrated.
Alkalis are soapy to the touch, and can cause burns.
Alkaline solutions have a pH greater than 7.

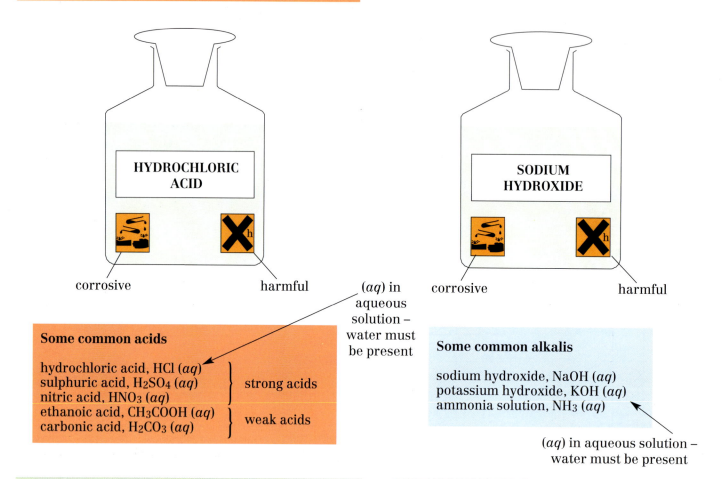

HYDROCHLORIC ACID

corrosive   harmful

(*aq*) in aqueous solution – water must be present

SODIUM HYDROXIDE

corrosive   harmful

Some common acids

hydrochloric acid, HCl (*aq*)
sulphuric acid, H_2SO_4 (*aq*)   } strong acids
nitric acid, HNO_3 (*aq*)
ethanoic acid, CH_3COOH (*aq*)   } weak acids
carbonic acid, H_2CO_3 (*aq*)

Some common alkalis

sodium hydroxide, NaOH (*aq*)
potassium hydroxide, KOH (*aq*)
ammonia solution, NH_3 (*aq*)

(*aq*) in aqueous solution – water must be present

The pH scale shows how acidic or alkaline a solution is. Universal indicator has a range of colours which can be matched against the pH scale.

Bases neutralise acids. Alkalis are **soluble** bases.

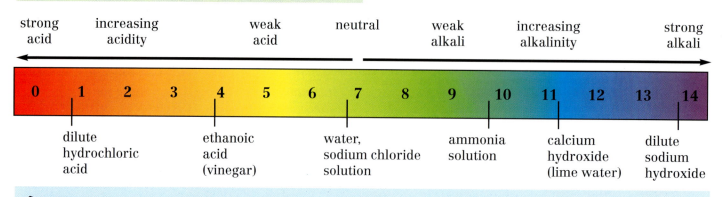

| strong acid | increasing acidity | | weak acid | neutral | weak alkali | increasing alkalinity | | strong alkali |

0 1 2 3 4 5 6 7 8 9 10 11 12 13 14

dilute hydrochloric acid

ethanoic acid (vinegar)

water, sodium chloride solution

ammonia solution

calcium hydroxide (lime water)

dilute sodium hydroxide

Questions

1. What are the chemical formulae of hydrochloric acid, sulphuric acid, and nitric acid?
2. Name two weak acids.
3. What are bases?
4. What are soluble bases called?

5. What colour is litmus solution in:
   (a) acids   (b) alkalis?
6. What do the symbols and tell you about a chemical?
7. *What safety precautions should be taken when working with acids and alkalis?*

ACIDS, BASES AND SALTS   Neutralisation I — Acids and alkalis

Neutralisation is a reaction between an **acid** and a **base** to give a **salt + water**.
Alkalis are **soluble** bases.

(A more detailed account of acid/alkali neutralisation is given on page 89.)

Experiment   Demonstrating neutralisation

See also pages 93, 102 and 103.

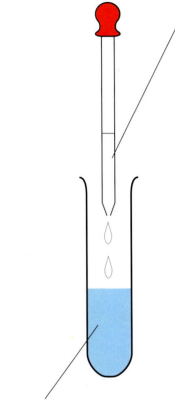

Dilute hydrochloric acid is added carefully until the point when one more drop of acid will change the indicator from blue to red – at this point, the solution is neutral.

ACID	+	ALKALI		SALT	+	WATER
hydrochloric acid	+	sodium hydroxide		sodium chloride	+	water
HCl (*aq*)	+	NaOH (*aq*)		NaCl (*aq*)	+	H_2O (*l*)

Sodium chloride is a salt –
it contains sodium from the alkali and chloride from the acid. Using the same method, another salt could be prepared, e.g.:

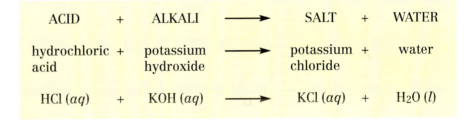

ACID	+	ALKALI		SALT	+	WATER
hydrochloric acid	+	potassium hydroxide		potassium chloride	+	water
HCl (*aq*)	+	KOH (*aq*)		KCl (*aq*)	+	H_2O (*l*)

Potassium chloride is a salt –
it contains potassium from the alkali and chloride from the acid.

Sodium hydroxide solution + a few drops of litmus solution – the solution is alkaline and turns the indicator blue.

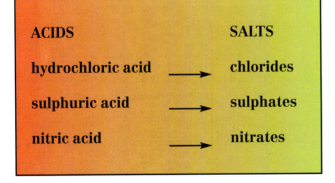

ACIDS		SALTS
hydrochloric acid	→	chlorides
sulphuric acid	→	sulphates
nitric acid	→	nitrates

Questions
1. What does the word 'neutralisation' mean?
2. Which salts are given by:
   (a) hydrochloric acid
   (b) sulphuric acid
   (c) nitric acid?
3. When a salt is prepared by acid/alkali neutralisation, where does the metal in the salt come from?
4. Is sodium chloride solution acidic, alkaline or neutral?
5. Which acid and alkali would you use to prepare the following salts:
   (a) sodium chloride   (b) potassium chloride
   (c) sodium sulphate   (d) potassium nitrate?
6. *Write word equations for the reactions in question 5.*
7. *What do the state symbols (aq) and (l) mean?*

ACIDS, BASES AND SALTS Neutralisation II — Acids and insoluble bases

KS3

Acids can be neutralised by insoluble bases, e.g. copper oxide, zinc oxide. This reaction can be used to prepare salts.

Experiment   Preparation of copper sulphate crystals

See also page 21.

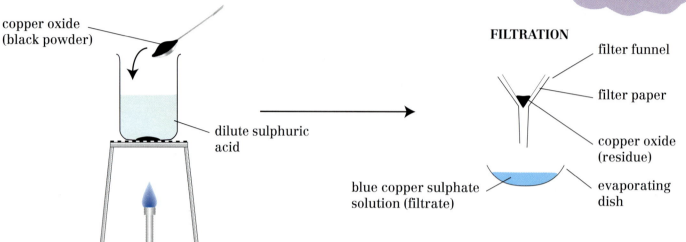

copper oxide (black powder)

dilute sulphuric acid

FILTRATION

filter funnel

filter paper

copper oxide (residue)

blue copper sulphate solution (filtrate)

evaporating dish

Copper oxide is added until it is in excess – i.e. there is some unreacted copper oxide in the mixture. This ensures that all the acid is used up.

The copper oxide reacts with the acid to give a blue solution.

ACID	+	INSOLUBLE BASE	→	SALT	+	WATER
sulphuric acid	+	copper oxide	→	copper sulphate	+	water
$H_2SO_4 \, (aq)$	+	$CuO \, (s)$	→	$CuSO_4 \, (aq)$	+	$H_2O \, (l)$

EVAPORATION

evaporating dish

blue copper sulphate solution

The solution is heated to reduce the volume.

CRYSTALLIZATION

crystallizing dish

Copper sulphate solution is transferred to a crystallizing dish and left in a warm place to allow crystals to form as the water evaporates from the solution.

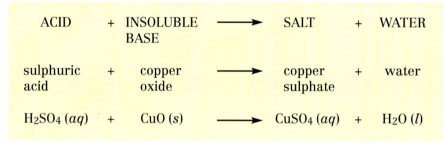

Questions

1. Why is the reaction between dilute sulphuric acid and copper oxide an example of a neutralisation reaction?
2. Why must excess copper oxide be used?
3. Why is the crystallizing dish left in a warm place?
4. *Which acids and bases would be used to prepare the following salts:*
   (a) copper chloride   *(b) copper nitrate*
   (c) zinc sulphate   *(d) zinc chloride?*
5. *Write word equations for each of the reactions in question 4.*
6. *What safety precautions must be taken when carrying out this experiment?*
7. *What does the state symbol (s) mean?*

When dilute acids react with carbonates, carbon dioxide is produced.

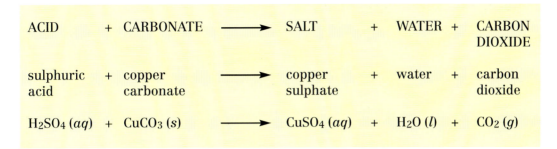

ACID	+	CARBONATE	⟶	SALT	+	WATER	+	CARBON DIOXIDE
sulphuric acid	+	copper carbonate	⟶	copper sulphate	+	water	+	carbon dioxide
H_2SO_4 (aq)	+	$CuCO_3$ (s)	⟶	$CuSO_4$ (aq)	+	H_2O (l)	+	CO_2 (g)

Experiment   Copper carbonate and sulphuric acid

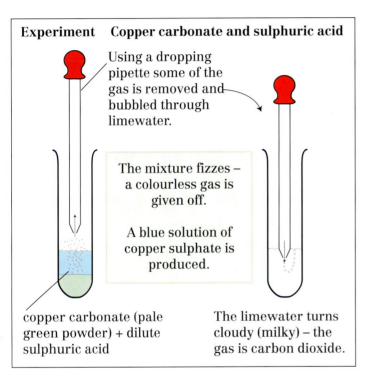

Using a dropping pipette some of the gas is removed and bubbled through limewater.

The mixture fizzes – a colourless gas is given off.

A blue solution of copper sulphate is produced.

copper carbonate (pale green powder) + dilute sulphuric acid

The limewater turns cloudy (milky) – the gas is carbon dioxide.

Experiment   Zinc carbonate and hydrochloric acid

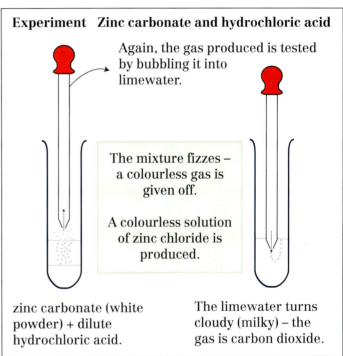

Again, the gas produced is tested by bubbling it into limewater.

The mixture fizzes – a colourless gas is given off.

A colourless solution of zinc chloride is produced.

zinc carbonate (white powder) + dilute hydrochloric acid.

The limewater turns cloudy (milky) – the gas is carbon dioxide.

Limewater turning cloudy is the chemical test for **carbon dioxide**.

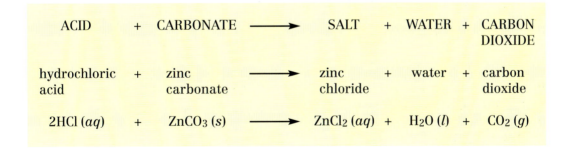

ACID	+	CARBONATE	⟶	SALT	+	WATER	+	CARBON DIOXIDE
hydrochloric acid	+	zinc carbonate	⟶	zinc chloride	+	water	+	carbon dioxide
2HCl (aq)	+	$ZnCO_3$ (s)	⟶	$ZnCl_2$ (aq)	+	H_2O (l)	+	CO_2 (g)

Production of carbon dioxide, when a dilute acid is added to a chemical, is a test for a **carbonate**.

Questions
1. Describe the chemical test for carbon dioxide.
2. Describe the chemical test for a carbonate.
3. *Write word equations for the following reactions:*
   (a) magnesium carbonate + hydrochloric acid ⟶
   (b) sodium carbonate + sulphuric acid ⟶
   (c) zinc carbonate + sulphuric acid ⟶
   (d) potassium carbonate + sulphuric acid ⟶
   (e) magnesium carbonate + sulphuric acid ⟶
   (f) copper carbonate + nitric acid ⟶
4. *What does the state symbol (g) mean?*

ACIDS, BASES AND SALTS Acids and metals

When dilute acids react with metals, hydrogen is produced.
This reaction can be used to prepare salts.

ACID	+	METAL	$\longrightarrow$	SALT	+ HYDROGEN
sulphuric acid	+	zinc	$\longrightarrow$	zinc sulphate	+ hydrogen
$H_2SO_4\ (aq)$	+	$Zn\ (s)$	$\longrightarrow$	$ZnSO_4\ (aq)$	+ $H_2\ (l)$

See also
page 73.

Experiment   Preparation of zinc sulphate crystals

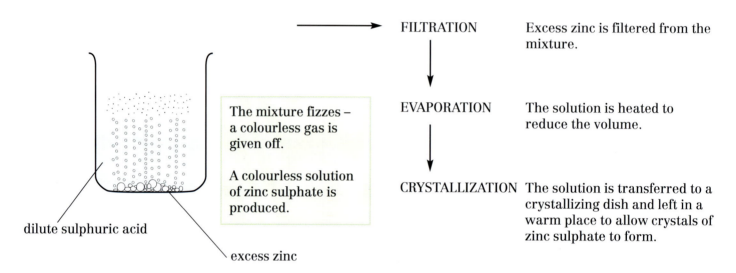

dilute sulphuric acid

excess zinc

The mixture fizzes –
a colourless gas is
given off.

A colourless solution
of zinc sulphate is
produced.

FILTRATION — Excess zinc is filtered from the mixture.

EVAPORATION — The solution is heated to reduce the volume.

CRYSTALLIZATION — The solution is transferred to a crystallizing dish and left in a warm place to allow crystals of zinc sulphate to form.

Questions

1. Why must excess zinc be added to the acid?
2. *From your knowledge of the Reactivity Series name:*
   (a) two other metals whose salts could be prepared by this method
   (b) two metals which would be dangerous to use in this experiment
   (c) two metals which would not react at all if this method were used.
3. *The zinc sulphate formed in this reaction is in aqueous solution (aq), but no water is produced in the reaction. Where does the water come from to make an aqueous solution?*
4. *The following table gives the names of some salts and some methods of preparation. For each salt place a tick or a cross in the spaces to show whether or not the salt could be prepared (safely) by the method shown.*

Salt	Method of preparation			
	Acid + alkali neutralisation	Acid + insoluble base neutralisation	Acid + carbonate	Acid + metal
sodium chloride				
calcium chloride				
magnesium sulphate				
zinc chloride				
copper sulphate				

ACIDS, BASES AND SALTS
Applications of neutralisation
Acids in the environment

Neutralisation reactions have many uses in everyday life.
The release of acidic gases and acidic solutions into the environment may cause serious problems.

Applications of neutralisation

1. Neutralisation of acidic soils and lakes polluted by acid rain

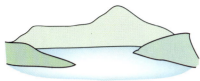

Acidic soils can be treated with:
calcium oxide (quicklime), CaO;
calcium hydroxide (slaked lime), Ca(OH)$_2$;
calcium carbonate (limestone), CaCO$_3$.

Acidic lakes can be treated with:
calcium hydroxide and calcium carbonate.

2. Indigestion
May be caused by the presence of too much acid in the stomach.

The excess acid can be neutralised by sodium hydrogen carbonate (bicarbonate of soda), NaHCO$_3$, or by an indigestion tablet.

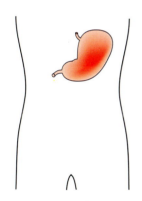

3. Treating factory waste
Many factories produce acidic waste, either as liquids or gases. These wastes are treated before being released into rivers or the atmosphere.

Powdered limestone is used to absorb acidic gases such as sulphur dioxide from the waste gases of power stations.

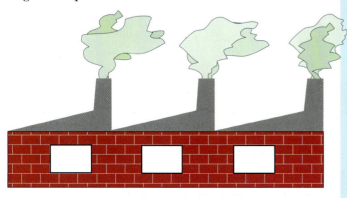

Acids in the environment

Pollution by acid rain
Burning fossil fuels such as coal and petrol produces gases which dissolve in water to form acids.

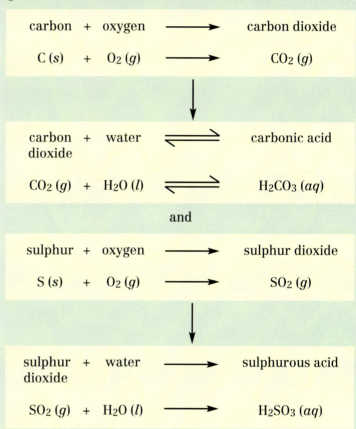

carbon + oxygen $\longrightarrow$ carbon dioxide

$C\,(s)$ + $O_2\,(g)$ $\longrightarrow$ $CO_2\,(g)$

carbon dioxide + water $\rightleftharpoons$ carbonic acid

$CO_2\,(g)$ + $H_2O\,(l)$ $\rightleftharpoons$ $H_2CO_3\,(aq)$

and

sulphur + oxygen $\longrightarrow$ sulphur dioxide

$S\,(s)$ + $O_2\,(g)$ $\longrightarrow$ $SO_2\,(g)$

sulphur dioxide + water $\longrightarrow$ sulphurous acid

$SO_2\,(g)$ + $H_2O\,(l)$ $\longrightarrow$ $H_2SO_3\,(aq)$

Acid rain causes damage to:
- limestone buildings, marble statues
- concrete and cement
- metals in bridges, car bodies
- trees and plant and animal life in rivers and lakes.

Questions
1. Give four problems caused by acid rain.
2. Name a chemical substance which could be used to treat indigestion caused by too much acid in the stomach.
3. Name two acidic gases and for each gas give the name of the acid formed when it dissolves in water.
4. Write word equations for the reactions between dilute hydrochloric acid and:
   (a) calcium oxide
   (b) calcium hydroxide
   (c) calcium carbonate.
5. Ant and nettle stings contain methanoic acid. Suggest an everyday substance that might be used to treat them.
6. Wasp stings are alkaline. Suggest an everyday substance that might be used to treat a wasp sting.

REVIEW QUESTIONS   Patterns of behaviour I

1. Name a metal which:
   (a) burns easily in oxygen with a bright white flame to produce a white powder
   (b) does not burn in oxygen, but becomes coated in a black powder when heated in air or oxygen
   (c) can be used to prevent the rusting of iron
   (d) reacts vigorously with dilute sulphuric acid
   (e) reacts violently with dilute sulphuric acid
   (f) reacts slowly with cold water, but reacts more vigorously with steam
   (g) removes oxygen from copper oxide
   (h) removes oxygen from iron oxide
   (i) displaces copper from copper sulphate solution
   (j) displaces silver from silver nitrate solution.

2. Complete the following diagram, which shows some of the reactions of zinc, by giving the names of the substances formed.

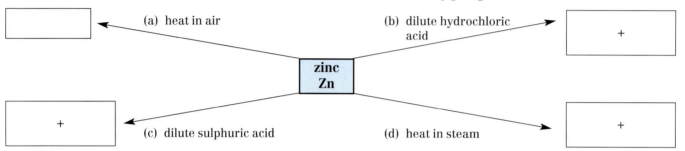

For each of the reactions (a) to (d) write a word equation.

3. Complete the following diagram, which shows some of the reactions of dilute sulphuric acid, by giving details of reactions which would give the products shown.

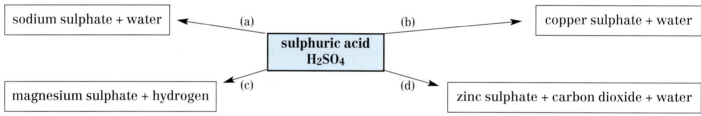

For each of the reactions (a) to (d) write a word equation.

4. On the following scale, show how acidity and alkalinity are associated with different pH values. Show also the colours that universal indicator would go in a strong acid, a weak acid, a neutral solution, a weak alkali, and a strong alkali.

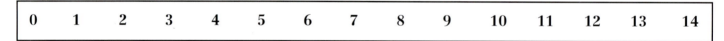

5. Complete each of the following word equations:
   (a) sodium hydroxide + nitric acid $\longrightarrow$
   (b) potassium hydroxide + hydrochloric acid $\longrightarrow$
   (c) potassium carbonate + hydrochloric acid $\longrightarrow$
   (d) copper carbonate + sulphuric acid $\longrightarrow$
   (e) zinc oxide + hydrochloric acid $\longrightarrow$
   (f) magnesium + silver nitrate $\longrightarrow$

6. Choose chemicals from the list below to illustrate each of the following types of reaction and give word equations for these reactions. Each chemical may be used once, more than once or not at all.

   (a) neutralisation using an acid and an alkali
   (b) neutralisation using an acid and an insoluble base
   (c) oxidation of a metal
   (d) reduction of a metal oxide
   (e) displacement of hydrogen from a dilute acid
   (f) displacement of a metal from a solution of one of its salts
   (g) an exothermic reaction.

calcium	sulphuric acid
copper	hydrochloric acid
iron	water
lead	copper oxide
magnesium	copper sulphate
silver	sodium hydroxide
sodium	zinc chloride
zinc	magnesium oxide

FORMULAE AND EQUATIONS II   Symbol equations I

Symbol equations must **balance**.
The number of atoms of each type must be the **same** on **both sides** of the equation.

NUMBERS AND FORMULAE

A number in **front of the formula** multiplies each atom in the formula

e.g.	2HCl	=	2 molecules of hydrochloric acid
		=	2 atoms of hydrogen and 2 atoms of chlorine.

A small number written **after a symbol in a formula** multiplies that atom within the formula

e.g.	H_2SO_4	=	sulphuric acid
		=	2 atoms of hydrogen, 1 atom of sulphur and 4 atoms of oxygen.

A small number **following a bracket** multiplies each of the atoms within the bracket

e.g.	$Ca(NO_3)_2$	=	calcium nitrate
		=	1 atom of calcium, 2 atoms of nitrogen and 6 atoms of oxygen.

HOW TO WRITE AN EQUATION

Example 1   **Neutralisation of sulphuric acid by sodium hydroxide:**

1. **Word equation**   sulphuric acid + sodium hydroxide ⟶ sodium sulphate + water

2. **Put in correct formulae** H_2SO_4 + NaOH ⟶ Na_2SO_4 + H_2O

3. **Check that each type of atom is balanced** H_2SO_4 + **2**NaOH ⟶ Na_2SO_4 + **2**H_2O

4. **Add state symbols** H_2SO_4 *(aq)* + **2**NaOH *(aq)* ⟶ Na_2SO_4 *(aq)* + **2**H_2O *(l)*

HOW TO WRITE AN EQUATION

Example 2   **Neutralisation of hydrochloric acid by copper oxide:**

1. **Word equation**   hydrochloric acid + copper oxide ⟶ copper chloride + water

2. **Put in correct formulae**   HCl + CuO ⟶ $CuCl_2$ + H_2O

3. **Check that each type of atom is balanced**   **2**HCl + CuO ⟶ $CuCl_2$ + H_2O

4. **Add state symbols**   **2**HCl *(aq)* + CuO *(s)* ⟶ $CuCl_2$ *(aq)* + H_2O *(l)*

IMPORTANT You cannot balance equations unless you are able to **write formulae correctly**.

Questions

Name each of the following compounds and state how many atoms of each type there are altogether in these formulae:

(a) NaCl
(b) FeS
(c) $CuCO_3$
(d) $ZnSO_4$
(e) Fe_2O_3
(f) $2CO_2$
(g) 2MgO
(h) $3H_2O$
(i) 2KOH
(j) 2PbO

FORMULAE AND EQUATIONS II Symbol equations II

HOW TO WRITE AN EQUATION

Example 3 Displacement of hydrogen from hydrochloric acid by magnesium:

1. Word equation	hydrochloric acid	+ magnesium	$\longrightarrow$	magnesium chloride	+ hydrogen	
2. Put in correct formulae	HCl	+ Mg	$\longrightarrow$	$MgCl_2$	+ H_2	
3. Check that each type of atom is balanced	2HCl	+ Mg	$\longrightarrow$	$MgCl_2$	+ H_2	
4. Add state symbols	2HCl (aq)	+ Mg (s)	$\longrightarrow$	$MgCl_2$ (aq)	+ H_2 (g)	

HOW TO WRITE AN EQUATION

Example 4 Displacement of copper from copper sulphate solution by a more reactive metal, magnesium:

1. Word equation	copper sulphate	+ magnesium	$\longrightarrow$	magnesium sulphate	+ copper
2. Put in correct formulae	$CuSO_4$	+ Mg	$\longrightarrow$	$MgSO_4$	+ Cu
3. Check that each type of atom is balanced	**IN THIS REACTION EACH TYPE OF ATOM IS ALREADY BALANCED**				
4. Add state symbols	$CuSO_4$ (aq)	+ Mg (s)	$\longrightarrow$	$MgSO_4$ (aq)	+ Cu (s)

Questions

1. For the following reactions, a word equation and an unbalanced symbol equation are given. Copy each of the equations and then complete the symbol equations by balancing each type of atom and adding state symbols.

(a) The reaction between hydrochloric acid and copper carbonate:

hydrochloric acid + copper carbonate $\longrightarrow$ copper chloride + water + carbon dioxide

HCl + $CuCO_3$ $\longrightarrow$ $CuCl_2$ + H_2O + CO_2

(b) The displacement of hydrogen from hydrochloric acid by zinc:

hydrochloric acid + zinc $\longrightarrow$ zinc chloride + hydrogen

HCl + Zn $\longrightarrow$ $ZnCl_2$ + H_2

(c) The reaction between sodium and water:

sodium + water $\longrightarrow$ sodium hydroxide + hydrogen

Na + H_2O $\longrightarrow$ NaOH + H_2

(d) The displacement of silver from silver nitrate by copper:

copper + silver nitrate $\longrightarrow$ copper nitrate + silver

Cu + $AgNO_3$ $\longrightarrow$ $Cu(NO_3)_2$ + Ag

2. Complete each of the following word equations and then proceed through the four stages to produce balanced symbol equations. Remember that some of these will balance immediately when you put in the symbols or formulae.

(a) sodium hydroxide + hydrochloric acid
(b) potassium hydroxide + sulphuric acid
(c) zinc oxide + hydrochloric acid
(d) zinc + sulphuric acid
(e) sulphuric acid + copper carbonate
(f) magnesium + oxygen
(g) sodium + oxygen
(h) potassium + water
(i) magnesium + water (steam)
(j) iron + copper oxide
(k) magnesium + zinc sulphate
(l) zinc + copper chloride

PERIODIC TABLE II Group trends in groups 1 and 7

Elements in the same group have similar properties. Within a particular group, there are trends in the physical and chemical properties.

Physical properties – characteristics of substances such as appearance, state [(s), (l), (g)], density, melting point and boiling point, solubility, whether or not they conduct heat and electricity.

Chemical properties – how a substance reacts in the environment of other chemicals, e.g. oxygen, water.

GROUP 1 ALKALI METALS

PHYSICAL PROPERTIES

- They are **metals**.

- They are **soft** and can be cut with a knife to give a **shiny surface**.

- They have **relatively low melting points** and **boiling points**.

- They have **low density** – lithium, sodium, and potassium float on water.

- They are **good conductors of heat and electricity**.

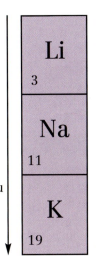

Melting points and boiling points **decrease** as you go down the group.

Density **decreases** as you go down the group.

| Li 3 |
| Na 11 |
| K 19 |

Reactivity **increases** as you go down the group.

CHEMICAL PROPERTIES

Summary

Details of these reactions can be found on pages 14, 74, 92 and 94.

- Alkali metals **react with water** to give hydroxides and hydrogen.

- Alkali metals **react with non-metals** to give ionic compounds.

GROUP 7 HALOGENS

PHYSICAL PROPERTIES

- They are **non-metals**.

- They have **coloured vapours**.

- They have **low melting points** and **boiling points**.

At room temperature:
  fluorine is a pale yellowish-green gas;
  chlorine is a greenish-yellow gas;
  bromine is a dark red-orange liquid;
  iodine is a dark grey crystalline solid.

- They are **poor conductors of heat and electricity** even when solid.

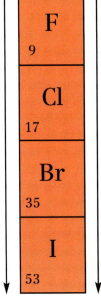

Melting points and boiling points **increase** as you go down the group.

Density **increases** as you go down the group.

| F 9 |
| Cl 17 |
| Br 35 |
| I 53 |

Reactivity **decreases** as you go down the group.

CHEMICAL PROPERTIES

Summary

Details of these reactions can be found on pages 14, 97 and 98.

- Halogens **react with metals** to give ionic compounds called salts.

- Halogens **combine with hydrogen** to give colourless gases.

PERIODIC TABLE II   Group trends in group 0

GROUP 0 NOBLE GASES

PHYSICAL PROPERTIES

- They are **non-metals**.

- At room temperature they are **colourless gases**.

- They have **low melting points** and **boiling points**.

- They are **poor conductors of heat and electricity**.

Melting points and boiling points **increase** as you go down the group.

Density **increases** as you go down the group.

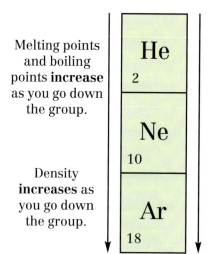

He
2
Ne
10
Ar
18

CHEMICAL PROPERTIES

- Noble gases are **unreactive**.

Neon lights
A high voltage is applied to the gas. Electrical energy is absorbed and light is emitted.

Questions

1. Name the members of Group 1 in order of increasing atomic number.
2. Name the members of Group 7 in order of increasing atomic number.
3. What is the trend in melting points and boiling points:
   (a) as you go down Group 1
   (b) as you go down Group 7?
4. What is the trend in density:
   (a) as you go down Group 1
   (b) as you go down Group 7?
5. What is the trend in reactivity:
   (a) as you go down Group 1
   (b) as you go down Group 7?
6. (a) What are the metallic properties of the alkali metals?
   (b) Are these properties physical or chemical properties?
7. (a) What are the non-metallic properties of the halogens?
   (b) Are these properties physical or chemical properties?
8. In what ways are the noble gases:
   (a) similar to the halogens in their **physical** properties
   (b) different from the halogens in their **physical** properties?
9. In what way are the noble gases different from the halogens in their **chemical** properties?
10. *The element below potassium in group 1 is rubidium. What would you expect the **physical** properties of rubidium to be? Explain your answer.*

PERIODIC TABLE II   Explaining group trends

In chemical reactions, an electron or electrons from the outer energy level (shell) is/are gained, lost or shared.
The reactivity of elements depends on how easily this happens.

GROUP 1   ALKALI METALS

In chemical reactions, alkali metals **lose** the **outer** electron.
Electrons are held in place in their electron shells by electrostatic attraction between the positive nucleus and negative electrons.

The outer electron is **lost** more easily if the metal has a larger atom because
(a) the **greater** the **distance** from the nucleus, the weaker the attraction of the nucleus for the electron.
(b) there is **more shielding** of the outer electron from the positively charged nucleus by electrons in the inner shells. The inner shells of negative electrons weaken the attraction of the nucleus for the outer electron.

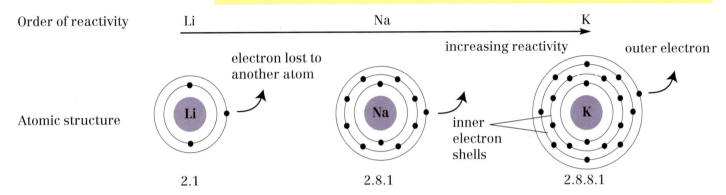

Order of reactivity

Atomic structure

GROUP 7   HALOGENS

In chemical reactions, halogens **gain** an electron **either** by an electron being transferred from a metal (ionic bonding) **or** by an electron being shared with a non-metal (covalent bonding).

An electron is **gained** more easily if the halogen has a smaller atom because
(a) the **smaller** the **distance** from the nucleus, the stronger the attraction of the nucleus for the electron.
(b) there is **less shielding** of the positively charged nucleus by electrons in the inner shell. Fewer inner shells of negative electrons mean that the attraction of the positive nucleus for an electron is stronger.

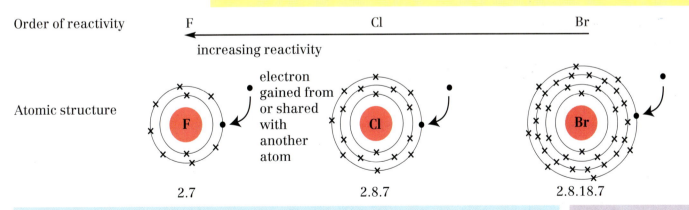

Order of reactivity

Atomic structure

Questions

1. What is the order of reactivity of the alkali metals (most reactive to least reactive)?
2. What is the order of reactivity of the halogens (most reactive to least reactive)?
3. Which shell of electrons (occupied energy levels) is involved in chemical reactions?
4. Explain why the reactivity of alkali metals increases as the atom becomes larger.
5. Explain why the reactivity of halogens increases as the atom becomes smaller.
6. *The element below potassium in Group 1 is rubidium. Would you expect it to be more or less reactive than potassium? Explain your answer.*
7. *The element below bromine in Group 7 is iodine. Would you expect it to be more or less reactive than bromine? Explain your answer.*

REMEMBER

When explaining patterns of reactivity, the key words are **distance** and **shielding**.

PERIODIC TABLE II   Period trends

In each period a particular energy level (electron shell) is gradually filled up with electrons. In period 3, the third energy level is filled.

On this page mass numbers of elements have been omitted.

Group																		0
1	2											3	4	5	6	7		He Helium 2

H
Hydrogen
1

transition metals

Group 1	2											3	4	5	6	7	0	Period
																	He Helium 2	1
Li Lithium 3	Be Beryllium 4											B Boron 5	C Carbon 6	N Nitrogen 7	O Oxygen 8	F Fluorine 9	Ne Neon 10	2
Na Sodium 11	Mg Magnesium 12											Al Aluminium 13	Si Silicon 14	P Phosphorus 15	S Sulphur 16	Cl Chlorine 17	Ar Argon 18	3
K Potassium 19	Ca Calcium 20	Sc Scandium 21	Ti Titanium 22	V Vanadium 23	Cr Chromium 24	Mn Manganese 25	Fe Iron 26	Co Cobalt 27	Ni Nickel 28	Cu Copper 29	Zn Zinc 30	Ga Gallium 31	Ge Germanium 32	As Arsenic 33	Se Selenium 34	Br Bromine 35	Kr Krypton 36	4
Rb Rubidium 37	Sr Strontium 38	Y Yttrium 39	Zr Zirconium 40	Nb Niobium 41	Mo Molybdenum 42	Tc Technetium 43	Ru Ruthenium 44	Rh Rhodium 45	Pd Palladium 46	Ag Silver 47	Cd Cadmium 48	In Indium 49	Sn Tin 50	Sb Antimony 51	Te Tellurium 52	I Iodine 53	Xe Xenon 54	5

PERIOD 3

sodium	magnesium	aluminium	silicon	phosphorus	sulphur	chlorine	argon
Na	Mg	Al	Si	P	S	Cl	Ar

        metals   metalloid   non-metals

high reactivity ⟶ low reactivity ⟵ high reactivity ⟶ unreactive

low melting point and boiling point ⟵ high melting point and boiling point ⟶ low melting point and boiling point

A metalloid is an element which shows some properties of a metal and some of a non-metal, e.g. silicon.

REMEMBER In chemical reactions an electron or electrons from the outer shell are gained, lost or shared.

Questions

1. Name three metals in Period 3.
2. Name three non-metals in Period 3.
3. Name a metalloid in Period 3.
4. What is a metalloid?
5. What is the trend in reactivity of metals in Period 3?
6. What is the trend in reactivity of non-metals in Period 3?
7. What is the trend in **physical properties** of metals in Period 3?
8. What is the trend in **physical properties** of non-metals in Period 3?

9. *How many electrons must be lost by a magnesium atom to give a full outer shell?*
10. *How many electrons must be gained or shared by an oxygen atom to give a full outer shell?*
11. *Why do you think sulphur is less reactive than chlorine?*
12. *Sodium has a melting point of 98°C and magnesium has a melting point of 649°C. What does this tell you about the strength of the bonds between the atoms in each of these elements?*

NOBLE GASES   Noble gases

Noble Gases are found in very small amounts in the air.
They are unreactive colourless gases.
They are useful because they are unreactive.

Gas	Symbol	Boiling point/°C
helium	He	−269
neon	Ne	−246
argon	Ar	−186
krypton	Kr	−157
xenon	Xe	−108

The **density increases** as the **mass** of the atom **increases**.

Helium is much less dense than air and is used in air balloons and weather balloons.

Neon is less dense than air and is used in advertising signs (neon lights).

Group 0

| He Helium 2 |
| Ne Neon 10 |
| Ar Argon 18 |
| Kr Krypton 36 |
| Xe Xenon 54 |

Neon glows red when an electric current passes through it.

Argon is slightly more dense than air and is used in filament lamps, electrical discharge tubes, and in welding to provide an unreactive atmosphere.

argon

tungsten filament

Krypton is more dense than air and is used in lasers.

Xenon is a dense gas and is used in powerful lamps, e.g. in lighthouses.

xenon

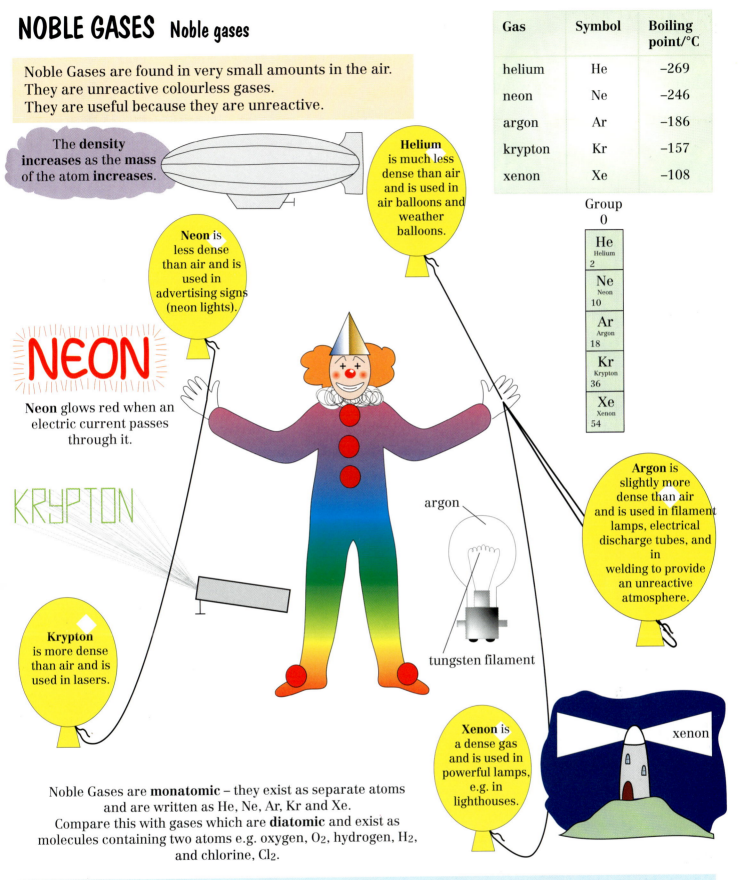

Noble Gases are **monatomic** – they exist as separate atoms and are written as He, Ne, Ar, Kr and Xe.
Compare this with gases which are **diatomic** and exist as molecules containing two atoms e.g. oxygen, O_2, hydrogen, H_2, and chlorine, Cl_2.

Questions

1. Which noble gases are less dense than air?
2. Which noble gases are more dense than air?
3. Why is helium used in balloons rather than hydrogen?
4. When tungsten is heated in air it reacts to form the oxide. Why does it not react inside a light bulb?
5. Which noble gas glows when an electric current passes through it?
6. Give one use each of krypton and xenon.
7. Of the five noble gases mentioned on this page, which has:
   (a) the lowest boiling point
   (b) the highest boiling point?
8. What is the trend in boiling points as you go down Group 0?
9. *In terms of electron arrangements explain why the noble gases are unreactive.*

ALKALI METALS Reactions with water

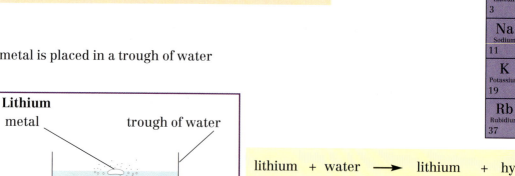

Alkali metals react with water to give hydroxides and hydrogen.
The hydroxides dissolve in water to give alkalis.

Experiment

A small piece of the metal is placed in a trough of water

Lithium

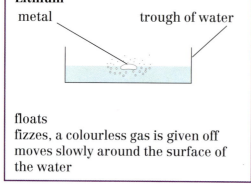

metal

trough of water

floats
fizzes, a colourless gas is given off
moves slowly around the surface of
the water

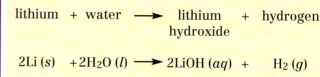

lithium + water $\longrightarrow$ lithium + hydrogen
hydroxide

$2Li\ (s)\ + 2H_2O\ (l) \longrightarrow 2LiOH\ (aq)\ +\ H_2\ (g)$

the reaction
becomes
increasingly
vigorous

Sodium

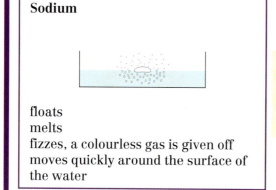

floats
melts
fizzes, a colourless gas is given off
moves quickly around the surface of
the water

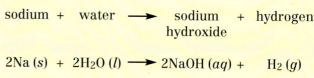

sodium + water $\longrightarrow$ sodium + hydrogen
hydroxide

$2Na\ (s)\ +\ 2H_2O\ (l) \longrightarrow 2NaOH\ (aq)\ +\ H_2\ (g)$

Potassium

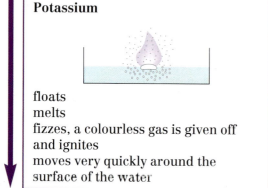

floats
melts
fizzes, a colourless gas is given off
and ignites
moves very quickly around the
surface of the water

potassium + water $\longrightarrow$ potassium + hydrogen
hydroxide

$2K\ (s)\ +\ 2H_2O\ (l) \longrightarrow 2KOH\ (aq)\ +\ H_2\ (g)$

If universal indicator is added to the solution at the end of the reaction, it turns violet.
Hydroxides dissolve in water to give alkalis.

Questions
1. Which gas is produced when alkali metals react
   with water?
2. Give the chemical formulae for
   (a) lithium hydroxide
   (b) sodium hydroxide
   (c) potassium hydroxide.
3. What do you notice about the way in which the
   symbol equations for these reactions balance?

4. How would you test a hydroxide solution to show
   that it is alkaline?
5. *Give the symbols for*
   (a) a lithium ion
   (b) a sodium ion
   (c) a potassium ion.
6. *Rubidium is below potassium in Group 1 of the*
   Periodic Table. How would you expect this metal to
   react with water? Explain your answer.

ALKALI METALS Alkalis and neutralisation

Ammonia dissolves in water to give an alkaline solution. It is a weak alkali. Ammonia solution can be neutralised to give **ammonium salts**.

All alkaline solutions contain hydroxide ions, OH⁻(aq).
Sodium hydroxide and potassium hydroxide are strong alkalis.

Neutralisation is a reaction between an acid and a base to give a salt and water.

Bases neutralise acids.
Alkalis are soluble bases.

$$ACID + BASE \longrightarrow SALT + WATER$$

In acid/alkali neutralisations the reaction is between H⁺ (aq) ions from the acid and OH⁻ (aq) ions from the alkali. These react together to form water. A salt is produced from the remaining ions.

$$H^+\ (aq) + OH^-\ (aq) \longrightarrow H_2O\ (l)$$

Phenolphthalein indicator

In acidic solutions colourless

In alkaline solutions pink

Titration is the method used in this experiment.
It measures the volumes of solutions which react together.

The type of salt produced depends on the acid used and on the metal in the alkali.

To prepare sodium sulphate you would use sulphuric acid and sodium hydroxide.

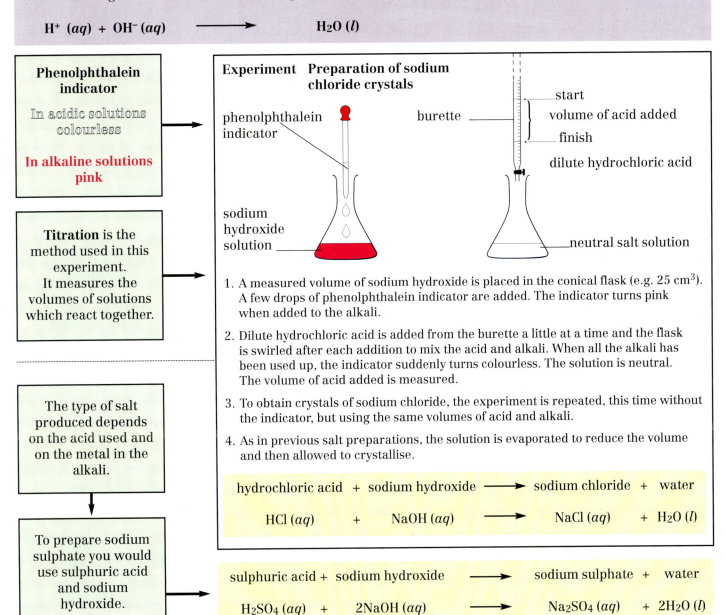

Experiment   Preparation of sodium chloride crystals

phenolphthalein indicator

sodium hydroxide solution

burette

start
volume of acid added
finish

dilute hydrochloric acid

neutral salt solution

1. A measured volume of sodium hydroxide is placed in the conical flask (e.g. 25 cm³). A few drops of phenolphthalein indicator are added. The indicator turns pink when added to the alkali.

2. Dilute hydrochloric acid is added from the burette a little at a time and the flask is swirled after each addition to mix the acid and alkali. When all the alkali has been used up, the indicator suddenly turns colourless. The solution is neutral. The volume of acid added is measured.

3. To obtain crystals of sodium chloride, the experiment is repeated, this time without the indicator, but using the same volumes of acid and alkali.

4. As in previous salt preparations, the solution is evaporated to reduce the volume and then allowed to crystallise.

hydrochloric acid + sodium hydroxide ⟶ sodium chloride + water

$$HCl\ (aq) \quad + \quad NaOH\ (aq) \quad \longrightarrow \quad NaCl\ (aq) \quad + \quad H_2O\ (l)$$

sulphuric acid + sodium hydroxide ⟶ sodium sulphate + water

$$H_2SO_4\ (aq) \quad + \quad 2NaOH\ (aq) \quad \longrightarrow \quad Na_2SO_4\ (aq) \quad + \quad 2H_2O\ (l)$$

Questions
1. Give an ionic equation for neutralisation.
2. What do you understand by the term 'titration'?
3. What colour is phenolphthalein in (a) acidic, and (b) alkaline solutions?
4. *If litmus solution had been used instead of phenolphthalein, what colour change would you have seen at neutralisation?*
5. *Name the salt which would be formed in each of the following neutralisation reactions:*
    (a) hydrochloric acid + potassium hydroxide (c) nitric acid + sodium hydroxide
    (b) sulphuric acid + potassium hydroxide   (d) nitric acid and ammonium hydroxide (ammonia solution).
6. Write word and symbol equations for each of the reactions in question 5.
7. What would be the pH of a strongly alkaline solution?
8. *What is the pH of ammonia solution?*

ALKALI METALS   Reactions with non-metals

Alkali metals react with non-metals to give ionic compounds.
The ionic compounds are solids which are soluble in water. The oxides dissolve in water to give alkaline solutions.

Experiment   Burning sodium in oxygen

burning
sodium
metal

oxygen

When sodium is heated in air, it melts and catches fire. Sodium burns more brightly in oxygen than in ordinary air. It has a yellow-orange flame.

In air sodium oxide, Na_2O, forms.
In oxygen sodium peroxide, Na_2O_2, also forms.

Potassium reacts in a similar way, burning with a lilac flame, but lithium, which burns with a red flame, produces only one oxide, Li_2O.

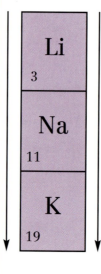

Li
3

Na
11

K
19

The reaction becomes increasingly vigorous

Experiment   Burning sodium in chlorine

burning
sodium
metal

chlorine

When burning sodium is placed in chlorine, it burns brightly and forms a white solid (**sodium chloride**).

When alkali metals react with halogen gases, the compounds formed are ionic salts containing halide ions.

Cl^-	chloride
Br^-	bromide
I^-	iodide

The equations below are for the formation of **oxides**.

The equations below are for the formation of **chlorides**.

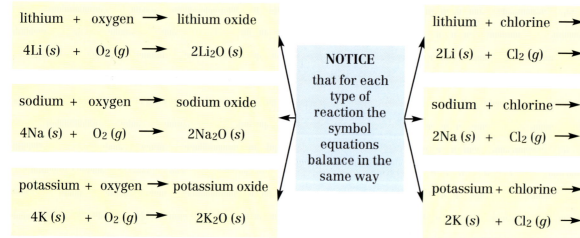

lithium + oxygen ⟶ lithium oxide

$$4Li\,(s) + O_2\,(g) \longrightarrow 2Li_2O\,(s)$$

sodium + oxygen ⟶ sodium oxide

$$4Na\,(s) + O_2\,(g) \longrightarrow 2Na_2O\,(s)$$

potassium + oxygen ⟶ potassium oxide

$$4K\,(s) + O_2\,(g) \longrightarrow 2K_2O\,(s)$$

NOTICE that for each type of reaction the symbol equations balance in the same way

lithium + chlorine ⟶ lithium chloride

$$2Li\,(s) + Cl_2\,(g) \longrightarrow 2LiCl\,(s)$$

sodium + chlorine ⟶ sodium chloride

$$2Na\,(s) + Cl_2\,(g) \longrightarrow 2NaCl\,(s)$$

potassium + chlorine ⟶ potassium chloride

$$2K\,(s) + Cl_2\,(g) \longrightarrow 2KCl\,(s)$$

Questions

1. What are the colours of the flames when lithium, sodium, and potassium burn in oxygen?
2. What do you notice about the way in which the symbol equations for these reactions balance?
3. Give the symbols for
   (a) a chloride ion   (b) a bromide ion   (c) a fluoride ion.
4. *Which elements are present in*
   (a) lithium oxide   *(c) potassium oxide*   *(e) potassium chloride*
   (b) sodium chloride   *(d) lithium chloride*   *(f) sodium oxide?*
5. *Fluorine is above chlorine in Group 7 of the Periodic Table. How would you expect sodium to react with fluorine? Give reasons for your answer.*

ALKALI METALS Sodium chloride

Sodium chloride is an important resource found in large quantities in seawater (brine) and in underground deposits.
Sodium chloride is an ionic compound containing Na^+ ions and Cl^- ions arranged in a crystal lattice. Solid sodium chloride does not conduct electricity because the ions are held together by strong forces of electrostatic attraction.
In order to conduct electricity, sodium chloride must be molten (melted) or in aqueous solution so that the ions are free to move.

The **electrolysis** of **sodium chloride solution** (brine) is an important industrial process.
The products are **hydrogen**, **chlorine**, and **sodium hydroxide**.

For electrolysis of solutions, **see pages 38 and 39.**

Uses of sodium chloride

(a) as rock salt to treat icy roads (it lowers the freezing point of water)
(b) in the food industry

Products of electrolysis of brine and their uses

Uses of chlorine, **see page 98.**

Hydrogen
- Haber Process for the production of ammonia
- hardening of vegetable oils to give margarine

Sodium hydroxide
- manufacture of soap, paper and ceramics

A membrane cell

What is happening

Sodium chloride solution contains four types of ion, as shown in the diagram (Na^+ and Cl^- from sodium chloride and H^+ and OH^- from water in the brine). When a current passes through the solution, the ions move towards the electrode with the opposite charge. At each electrode, one product is formed.

At the anode, chlorine bubbles off:

$$2Cl^- \ (aq) - 2e^- \longrightarrow Cl_2 \ (g)$$

At the cathode, hydrogen bubbles off:

$$2H^+ \ (aq) + 2e^- \longrightarrow H_2 \ (g)$$

Na^+ ions and OH^- ions remain behind and form **sodium hydroxide**.

Questions

1. Where is sodium chloride found naturally?
2. Which ions are present in sodium chloride? Give the symbols for these ions.
3. If sodium chloride **solution** is electrolysed, what will be the product
   (a) at the anode   (b) at the cathode?
4. Give two uses for each of the products of the electrolysis of brine.
5. Why does solid sodium chloride not conduct electricity?
6. What is the importance of the membrane in the 'membrane cell'?
7. *In the membrane cell, where do the hydrogen ions, H⁺, and hydroxide ions, OH⁻, come from?*
8. *If **molten** sodium chloride is electrolysed, what will be the product:*
   (a) at the anode   (b) at the cathode?

HALOGENS The halogens

The halogens are the elements of group 7.

- The halogens have **seven** electrons in the outer shell.
- The halogens are **non-metals**.
- The halogens form simple diatomic covalent molecules – F_2, Cl_2, Br_2, I_2.

Chlorine atoms

A simple molecular structure of two atoms combined.

The properties of metals and **non-metals** are given on pages 9 and 10.

7

Fluorine –
a pale yellowish-green **gas**

Chlorine –
a greenish-yellow **gas**

Bromine –
a **dark** red-orange **liquid**

Iodine –
a grey **solid**

| F |
| Fluorine 9 |
| Cl |
| Chlorine 17 |
| Br |
| Bromine 35 |
| I |
| Iodine 53 |
| At |
| Astatine 85 |

The smell of chlorine is commonly associated with bleaches and swimming pools, but too much chlorine can be harmful because it is a poisonous gas. The bleaching properties can be used as a test for chlorine because it turns a piece of moist indicator paper **white**.

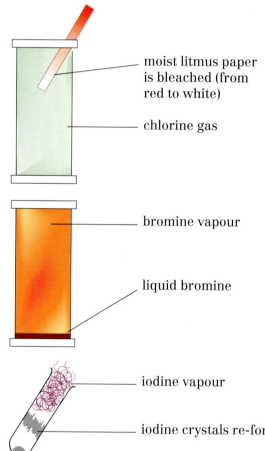

moist litmus paper is bleached (from red to white)

chlorine gas

Great care should be taken with liquid bromine. It causes serious damage to skin and readily turns into an orange gas which is very poisonous. It should only be used in a fume cupboard, and containers of bromine should be well sealed.

bromine vapour

liquid bromine

A crystal of iodine will give off a poisonous, purple vapour when heated. If crystals of iodine are left in an open container they slowly vaporise and gradually disappear. **Solid** iodine **sublimes**, which means that it does not melt but **turns directly into a gas**.

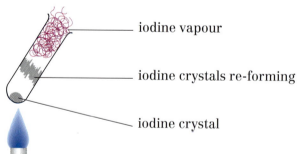

iodine vapour

iodine crystals re-forming

iodine crystal

- Halogens form ionic bonds with metals.

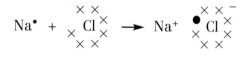

See page 32.

- Halogens form covalent bonds with non-metals.

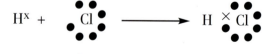

Questions
1. Describe chlorine.
2. What changes of state occur down group 7 from chlorine to iodine?
3. *There is a fifth halogen called astatine which is below iodine. Predict its physical state and colour.*
4. *Why are the halogens in group 7?*
5. Describe a test for chlorine.
6. What happens to liquid bromine left in an open container?
7. What is unusual about the effect of heat on iodine?
8. What are the bonding, structure and formula of chlorine gas?

HALOGENS Reactions of halogens

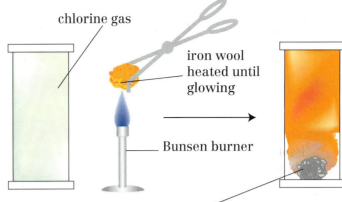

Chlorine reacts with metals.

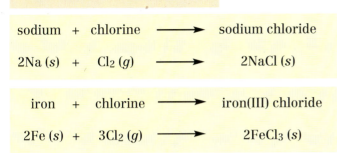

sodium + chlorine $\longrightarrow$ sodium chloride

$2Na\ (s)\ +\ Cl_2\ (g)\ \longrightarrow\ 2NaCl\ (s)$

iron + chlorine $\longrightarrow$ iron(III) chloride

$2Fe\ (s)\ +\ 3Cl_2\ (g)\ \longrightarrow\ 2FeCl_3\ (s)$

- The halogens react with metals by gaining an electron to complete the energy level (outer shell).
- Their reactivity depends on how strongly they can attract and hold the new electron.
- Fluorine is the most reactive because it attracts and holds a new electron most strongly.

$F\ 2.7$ gains e^- $\longrightarrow$ $F^-\ 2.8$

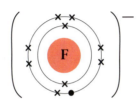

The fluoride ion has the new electron in the second shell.
It is closer to the nucleus and only shielded by the first shell.

$Cl\ 2.8.7$ gains e^- $\longrightarrow$ $Cl^-\ 2.8.8$

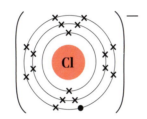

The chloride ion has the new electron in the third shell.
It is further from the nucleus and shielded by the first and second shells.

- There is a pattern down the group.
- As the outer shell is further from the nucleus the element becomes less reactive.
- Chlorine is more reactive than bromine and iodine but not as reactive as fluorine.

most reactive	fluorine	F_2
	chlorine	Cl_2
↓	bromine	Br_2
least reactive	iodine	I_2

Chlorine displaces bromine and iodine.

- Although chlorine, bromine, and iodine are coloured most of their compounds are not.

Bromine $\longrightarrow$ Sodium bromide $\longrightarrow$ Sodium bromide solution
red-orange liquid   white solid   colourless solution

- If chlorine gas reacts with sodium bromide solution the orange colour of bromine appears in the liquid.
- Chlorine is more reactive than bromine and therefore displaces it from its compounds.
- Both chlorine and bromine will displace iodine from sodium iodide solution because they are more reactive than iodine.

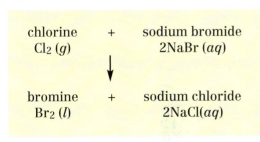

chlorine + sodium bromide
$Cl_2\ (g)$ $2NaBr\ (aq)$

↓

bromine + sodium chloride
$Br_2\ (l)$ $2NaCl(aq)$

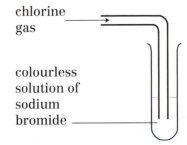

chlorine gas

colourless solution of sodium bromide

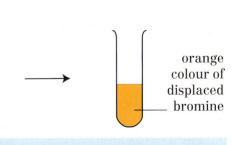

orange colour of displaced bromine

Questions

1. Write an equation for the reaction between sodium and chlorine.

2. What happens when chlorine reacts with sodium bromide solution?

HALOGENS Uses of halogens and their compounds

- Chlorine reacts directly with hydrogen to form the covalent compound **hydrogen chloride**.

- Hydrogen chloride is a colourless gas which dissolves in water to form **hydrochloric acid**.

- Hydrogen chloride, hydrogen bromide and hydrogen iodide all dissolve in water to produce **acids** because they form $H^+(aq)$ ions.

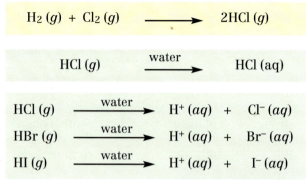

$$H_2 (g) + Cl_2 (g) \longrightarrow 2HCl (g)$$

$$HCl (g) \xrightarrow{\text{water}} HCl (aq)$$

$$HCl (g) \xrightarrow{\text{water}} H^+ (aq) + Cl^- (aq)$$
$$HBr (g) \xrightarrow{\text{water}} H^+ (aq) + Br^- (aq)$$
$$HI (g) \xrightarrow{\text{water}} H^+ (aq) + I^- (aq)$$

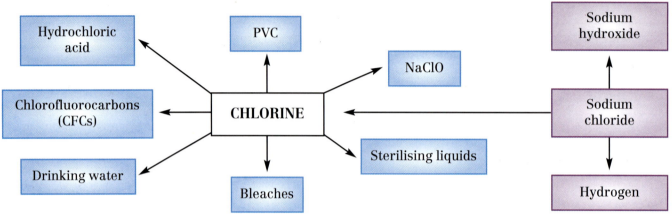

- Drinking water — Tap water is **chlorinated** to kill harmful bacteria so that it is fit to drink.
- Sterilising liquids — Contain **dilute** solutions of sodium chlorate(I), NaClO (sodium hypochlorite). Some also contain sodium chloride solution.
- Bleaches — Contain solutions of sodium chlorate (I). Some thick bleaches also contain sodium hydroxide, which helps dissolve grease.
- PVC — Poly(chloroethene) is a plastic used for insulating electric cables and for articles of clothing such as raincoats. It is also called poly(vinylchloride).
- Hydrochloric acid — Manufactured by burning hydrogen in chlorine to make hydrogen chloride.
- CFCs — Used in aerosols, fridges, and packaging foams but can damage the ozone layer.

The bleaching or sterilising action of sodium chlorate(I) occurs because it decomposes to produce reactive **atoms of oxygen**.

Normally the oxygen atoms combine to form $O_2 (g)$ but they may also attack bacteria, or react with dyes causing them to become colourless.

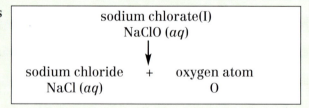

- **Iodine solution** (actually a solution of iodine and potassium iodide in water)
  may be used as an antiseptic to kill germs on cuts and grazes.
- Iodine solution may be used to test for the presence of starch.
- The presence of **sodium fluoride** in drinking water **helps to prevent tooth decay**.
- Sodium fluoride is naturally present in some sources of tap water and is added to others.
- Fluoride is also used in some toothpastes for the same reason.

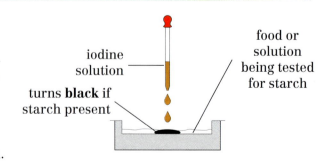

Questions

1. What is the common name for hydrogen chloride solution and why is it acidic?
2. In what form is iodine applied to cuts and why is it used?
3. State two uses of sodium chlorate(I) solution.
4. Why is tap water treated with chlorine and why is sodium fluoride added?

TRANSITION METALS Transition metals

Transition metals have properties suitable for structural and engineering work.

Sc	Ti	V	Cr	Mn	Fe	Co	Ni	Cu	Zn
Scandium	Titanium	Vanadium	Chromium	Manganese	Iron	Cobalt	Nickel	Copper	Zinc
21	22	23	24	25	26	27	28	29	30

Transition elements appear in the block between group 2 and group 3 in the Periodic Table beginning at scandium, Sc. These elements are all metals and their properties may be compared with the metals of groups 1, 2, and 3. Transition metals are generally less reactive, stronger, more dense, and have higher melting points, which makes them more useful for engineering work.

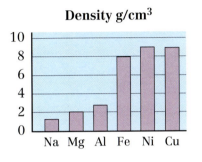

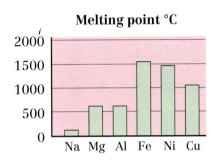

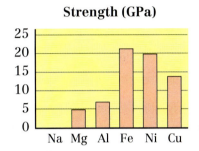

- Strength indicates the ability to resist forces acting on the metal.
- Definitions of strength account for different types of force such as stretching, bending or being compressed.
- Hardness and strength are different.
- Hard materials are those which do not scratch easily.

Some transition metals are used as **catalysts**. Iron is used in the Haber process for the manufacture of ammonia, and nickel is used in the manufacture of margarine.
More uses? **See page 125.**

Transition elements form compounds which have characteristic colours.

COPPER(II) SULPHATE	IRON(III) CHLORIDE	NICKEL(II) SULPHATE

Copper compounds are often blue or blue-green, and this may be seen on the domed roofs of some buildings where the copper cladding has weathered and turned green (verdigris).

Rust is a common example of an iron(III) compound with a typical orange/brown colour.

Coloured precipitates? **See page 104.**

The colours in glazes on pottery are often produced by transition metals (e.g. chromium oxide can give green, cobalt carbonate can produce blue).

Uses of copper? **See page 56.**

Uses of iron? **See page 55.**

Questions

1. Name three metals which are not transition metals and three which are.
2. Compare the densities and melting points of transition metals with those of other metals.
3. Why are transition metals used in structural and engineering work?
4. What is the purpose of iron in the manufacture of ammonia?
5. What are the characteristic colours of copper compounds and iron(III) compounds?

REVIEW QUESTIONS Patterns of behaviour II

1. Name each of the following compounds and state how many atoms of each type there are altogether in the following formulae:
   (a) KOH
   (b) $ZnSO_4$
   (c) $PbBr_2$
   (d) H_2CO_3
   (e) 2KCl
   (f) 2NaBr
   (g) $2HNO_3$
   (h) $2AgNO_3$
   (i) $3CH_4$
   (j) 2LiOH
   (k) $2NH_3$
   (l) $Ca(HCO_3)_2$
   (m) $(NH_4)_2SO_4$
   (n) $Zn(NO_3)_2$
   (o) $Mg(OH)_2$
   (p) $Cu(NO_3)_2$.

2. Give:
   (a) three reactions which are neutralisation reactions
   (b) two reactions which are oxidation reactions, where oxygen is added
   (c) three reactions which are displacement reactions.

3. For each of the reactions in question 2 proceed through the four stages to produce balanced symbol equations.

4. Balance the following symbol equations:
   (a) $CH_4 + O_2 \longrightarrow CO_2 + H_2O$
   (b) $H_2SO_4 + CuCO_3 \longrightarrow CuSO_4 + H_2O + CO_2u$
   (c) $CaCO_3 \xrightarrow{heat} CaO + CO_2$
   (d) $ZnO + HCl \longrightarrow ZnCl_2 + H_2O$
   (e) $Zn + CuSO_4 \longrightarrow ZnSO_4 + Cu$
   (f) $H_2 + Cl_2 \longrightarrow HCl$
   (g) $Al_2O_3 \longrightarrow Al + O_2$ (electrolysis)
   (h) $PbO + C \longrightarrow Pb + CO_2$

5. What is meant by the terms:
   (a) physical properties   (b) chemical properties?

6. (a) What name is given to the elements in group 1?
   (b) What do the elements of group 1 have in common with regard to the arrangement of their outer electrons?
   (c) Describe the appearance of lithium, sodium and potassium.
   (d) What is the trend as you go down group 1 in terms of:
       (i) density; (ii) melting point and boiling point; (iii) reactivity?
   (e) How does the size of the atoms account for the trends in reactivity of the elements in group 1?

7. (a) List the elements in period 3 of the Periodic Table.
   (b) Which elements in period 3 are highly reactive?
   (c) Which elements in period 3 have low melting points?
   (d) Which element in period 3 is the least reactive?

8. (a) Give three examples of monatomic gases.
   (b) Give three examples of diatomic gases.
   (c) Describe two uses of Noble Gases which depend on the fact that they are unreactive.

9. (a) Give the formulae of lithium hydroxide, sodium hydroxide, and potassium hydroxide.
   (b) Write word and symbol equations for the reactions of each of these alkalis with sulphuric acid.
   (c) What is meant by: (i) neutralisation; (ii) a salt?

10. (a) Name two non-metals which react with alkali metals.
    (b) What is meant by the term direct combination?
    (c) Give the formulae of: (i) lithium bromide; (ii) sodium bromide; (iii) potassium bromide.

11. (a) What are the main sources of sodium chloride?
    (b) Why is it an important resource?
    (c) What are the products of the electrolysis of:
        (i) molten sodium chloride; (ii) brine (sodium chloride solution)?
    (d) Why is the electrolysis of brine of economic importance?

12. (a) What name is given to the elements in group 7? What type of elements are they?
    (b) What do the elements of group 7 have in common with regard to the arrangement of their outer electrons?
    (c) Describe the physical state and appearance of chlorine, bromine, and iodine.

13. (a) Write equations for the reaction of sodium with bromine, and magnesium with chlorine.
    (b) What type of bond is most likely to be formed when a metal reacts with chlorine?
    (c) Write an equation for the reaction between hydrogen and chlorine.
    (d) Explain how hydrogen chloride becomes hydrochloric acid.
    (e) Why do hydrogen bromide and hydrogen iodide also produce acids?

14. (a) Write an equation for a displacement reaction involving chlorine and potassium bromide, including state symbols.
    (b) Explain why chlorine displaces iodine, and how bromine would react with solutions of sodium chloride and sodium iodide.
    (c) Describe how bromine water can be used to distinguish ethane from ethene.

15. (a) Name three transition metals and indicate in what ways they will be different from main group metals like sodium or magnesium.
    (b) State a characteristic property of compounds of transition metals.
    (c) Give uses of three named transition metals.
    (d) What is steel, and why are other transition metals added to some steels?

REACTIONS Types of reaction

Combination and combustion are examples of different types of reaction.

iron + sulphur $\longrightarrow$ iron sulphide

$Fe(s) + S(s) \longrightarrow FeS(s)$

methane + oxygen $\longrightarrow$ carbon dioxide + water

$CH_4(g) + 2O_2(g) \longrightarrow CO_2(g) + 2H_2O(l)$

COMBINATION
Iron and sulphur combine together.

COMBUSTION
Methane burns in oxygen.

COMBINATION
Elements combine to form a compound.

NEUTRALISATION
Acids are neutralised by an alkali or base.

DISPLACEMENT
One element takes the place of another in a compound.

PRECIPITATION
Substances in solution react to form an insoluble substance.

TYPES OF REACTION

COMBUSTION
A substance burns in air or oxygen.

REDUCTION
Oxygen is lost or hydrogen is gained or electrons are gained.

DECOMPOSITION
A compound splits up into simpler substances.

OXIDATION
Oxygen is gained or hydrogen is lost or electrons are lost.

Decomposition can be brought about by heat or a catalyst or both.

THERMAL DECOMPOSITION
Means that heat has been used to split up a compound.

CATALYTIC DECOMPOSITION
Means that a catalyst has been used to speed up the breakdown of a compound.

THERMAL/CATALYTIC CRACKING

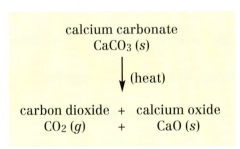

calcium carbonate
$CaCO_3 (s)$

↓ (heat)

carbon dioxide + calcium oxide
$CO_2 (g)$ + $CaO (s)$

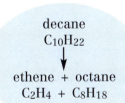

decane
$C_{10}H_{22}$

↓

ethene + octane
$C_2H_4 + C_8H_{18}$

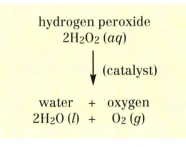

hydrogen peroxide
$2H_2O_2 (aq)$

↓ (catalyst)

water + oxygen
$2H_2O (l)$ + $O_2 (g)$

- **Cracking** is a **decomposition reaction** in which molecules containing long hydrocarbon chains are broken down into smaller molecules.
- **Thermal cracking** uses **heat** alone whereas **catalytic cracking** uses **heat** and a **catalyst**.
- The temperatures used range from 400°C to 800°C; with a catalyst less heat is needed.

Cracking.
See page 48.

Questions

1. Write a word and symbol equation for the reaction between sulphur and oxygen.
2. Explain why the reaction of sulphur in question 1 is oxidation, combustion, and combination.
3. What are combustion reactions?
4. Write an equation for the combination reaction between Mg and Cl_2.
5. Why is cracking a decomposition?
6. Name the two types of cracking and explain the difference.

REACTIONS Displacement and neutralisation

Displacement reactions usually take place in solution, and are those in which one element takes the place of another in a compound.

iron + copper sulphate ⟶ iron sulphate + copper Fe(s) + CuSO₄(aq) ⟶ FeSO₄(aq) + Cu(s)	A more reactive metal will displace a less reactive metal from solutions of its salts. Iron displaces the copper.

iron + copper sulphate ⟶ iron sulphate + copper

$Fe(s)$ + $CuSO_4(aq)$ ⟶ $FeSO_4(aq)$ + $Cu(s)$

> A more reactive metal will displace a less reactive metal from solutions of its salts. Iron displaces the copper.

chlorine + sodium bromide ⟶ sodium chloride + bromine

$Cl_2(g)$ + $2NaBr(aq)$ ⟶ $2NaCl(aq)$ + $Br_2(l)$

> Chlorine is more reactive than bromine and therefore displaces it from its compound sodium bromide.

zinc + sulphuric acid ⟶ zinc sulphate + hydrogen

$Zn(s)$ + $H_2SO_4(aq)$ ⟶ $ZnSO_4(aq)$ + $H_2(g)$

> This type of reaction is not always referred to as displacement, nevertheless the zinc does displace hydrogen from the acid.

Neutralisation reactions are those in which acids react with bases, including alkalis, to give a salt and water.

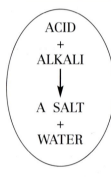

hydrochloric acid + sodium hydroxide

$HCl(aq)$ + $NaOH(aq)$

↓

sodium chloride + water

$NaCl(aq)$ + $H_2O(l)$

sulphuric acid + magnesium oxide

$H_2SO_4(aq)$ + $MgO(s)$

↓

magnesium sulphate + water

$MgSO_4(aq)$ + $H_2O(l)$

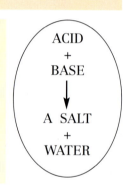

ammonia + sulphuric acid ⟶ ammonium sulphate

$2NH_3(aq)$ + $H_2SO_4(aq)$ ⟶ $(NH_4)_2SO_4(aq)$

> The reaction between ammonia and sulphuric acid is a neutralisation because ammonia solution is a weak alkali.

Strong acids may be 'neutralised' by carbonates but this might not be classified as neutralisation because a **salt and water** are not the **only** products.

hydrochloric acid + calcium carbonate ⟶ calcium chloride + carbon dioxide + water

$2HCl(aq)$ + $CaCO_3(s)$ ⟶ $CaCl_2(aq)$ + $CO_2(g)$ + $H_2O(l)$

Questions

1. Write an equation for the reaction between zinc and copper sulphate solution, state the type of reaction it is, and explain why silver will not react with copper sulphate.
2. Write an equation for the reaction between bromine and sodium iodide solution, and explain the reaction in terms of displacement.
3. Write an equation for the reaction between magnesium and hydrochloric acid.
4. Which two **types** of substance will neutralise acids?
5. Which two **types** of substance are produced during neutralisation?
6. Name the two types of salt produced by hydrochloric acid and sulphuric acid.
7. Give the formula and name of the acid formed when carbon dioxide mixes with water, and write an equation for the reaction of this acid with sodium hydroxide solution.
8. Why can sodium hydroxide be used to absorb carbon dioxide?

REACTIONS Redox reactions and ionic equations

Redox reactions are those which involve both oxidation and reduction.

- All reactions involving the gain and loss of oxygen or hydrogen or electrons are **redox** reactions.

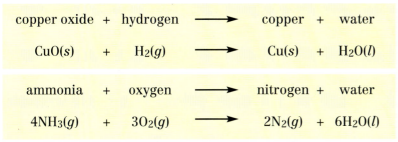

copper oxide + hydrogen ⟶ copper + water

CuO(*s*) + H₂(*g*) ⟶ Cu(*s*) + H₂O(*l*)

Copper oxide loses oxygen and is reduced. Hydrogen gains oxygen and is oxidised. Hydrogen is the **reducing agent**.

ammonia + oxygen ⟶ nitrogen + water

4NH₃(*g*) + 3O₂(*g*) ⟶ 2N₂(*g*) + 6H₂O(*l*)

Ammonia loses hydrogen and is oxidised. Oxygen gains hydrogen and is reduced. Oxygen is the **oxidising agent**.

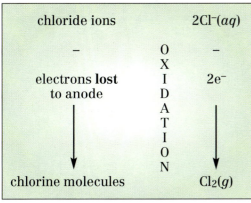

chloride ions .... 2Cl⁻(*aq*)

– ... O X I D A T I O N ... –

electrons **lost** to anode ... 2e⁻

chlorine molecules .... Cl₂(*g*)

During electrolysis, ions **lose electrons** at the anode (+) and **gain electrodes** at the cathode (–). This means that both oxidation and reduction are taking place.

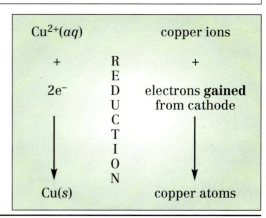

Cu²⁺(*aq*) .... copper ions

+ ... R E D U C T I O N ... +

2e⁻ ... electrons **gained** from cathode

Cu(*s*) .... copper atoms

Reactions which take place in aqueous solution often involve **ions**.

Oxidation Is Loss. Reduction Is Gain. **OIL RIG** (For electrons)

- Sometimes it is useful to show all the ions using a **full ionic equation**.
- **Essential ionic equations** show only those ions which are reacting.

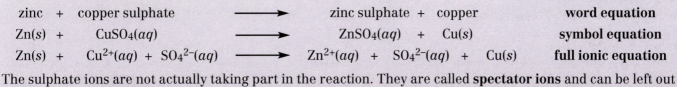

zinc + copper sulphate ⟶ zinc sulphate + copper	**word equation**	
Zn(*s*) + CuSO₄(*aq*) ⟶ ZnSO₄(*aq*) + Cu(*s*)	**symbol equation**	
Zn(*s*) + Cu²⁺(*aq*) + SO₄²⁻(*aq*) ⟶ Zn²⁺(*aq*) + SO₄²⁻(*aq*) + Cu(*s*)	**full ionic equation**	

The sulphate ions are not actually taking part in the reaction. They are called **spectator ions** and can be left out of the equation. The essential ionic equation includes only those atoms and ions which are reacting.

Zn(*s*) + Cu²⁺(*aq*) ⟶ Zn²⁺(*aq*) + Cu(*s*);   **essential ionic equation**

zinc + copper ions .... zinc ions + copper

- The reaction between zinc and copper sulphate solution is classified as a displacement reaction.
- The ionic equations show that it is also a redox reaction because electrons have been exchanged.
- The essential ionic equation shows that the sulphate ions are not involved, which means that zinc should react with any solution containing copper ions.

Displacement. **See page 77.**

hydrochloric acid + sodium hydroxide ⟶ sodium chloride + water   **word equation**

HCl(*aq*) + NaOH(*aq*) ⟶ NaCl(*aq*) + H₂O(*l*)   **symbol equation**

H⁺(*aq*) + Cl⁻(*aq*) + Na⁺(*aq*) + OH⁻(*aq*) ⟶ Na⁺(*aq*) + Cl⁻(*aq*) + H₂O(*l*)   **full ionic equation**

The ionic equation shows that for this neutralisation reaction the ions which are reacting are hydrogen ions, H⁺(aq), from the acid and hydroxide ions, OH⁻(aq), from the alkali. These ions react to produce neutral water. The essential ionic equation represents the neutralisation reaction between any acid and any alkali:

H⁺(*aq*) + OH⁻(*aq*) ⟶ H₂O(*l*)   **essential ionic equation for neutralisation**

Questions

1. Why is the reaction between hydrogen and copper oxide a redox reaction?
2. Why are the electrode reactions during electrolysis redox reactions?
3. Write a symbol equation and an ionic equation for a neutralisation reaction.

REACTIONS Precipitation reactions

In **precipitation** reactions an insoluble substance forms within a solution.

- An insoluble product cannot remain dissolved so forms a <u>s</u>olid which causes cloudiness.
- Precipitation reactions usually take place between two solutions.

silver nitrate + sodium chloride $\longrightarrow$ silver chloride + sodium nitrate	Silver chloride is not soluble in water so it appears as a white precipitate.
$AgNO_3(aq)$ + $NaCl(aq)$ $\longrightarrow$ $AgCl(s)$ + $NaNO_3(aq)$	

carbon dioxide + calcium hydroxide $\longrightarrow$ calcium carbonate + water	Carbon dioxide turns limewater cloudy because a white precipitate of insoluble calcium carbonate forms.
$CO_2(g)$ + $Ca(OH)_2(aq)$ $\longrightarrow$ $CaCO_3(s)$ + $H_2O(l)$	

- Some precipitation reactions are used as tests for **types** of compounds.

- **Silver nitrate** solution gives a **white precipitate** with solutions of **chlorides**.

- **Barium chloride** solution gives a **white precipitate** with solutions of **sulphates**.

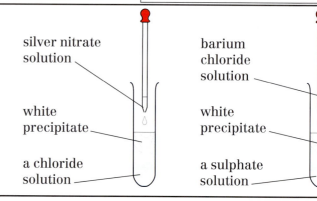

silver nitrate solution · white precipitate · a chloride solution · barium chloride solution · white precipitate · a sulphate solution

Copper sulphate solution reacts with sodium hydroxide solution.
The reaction produces a **blue precipitate**.
The blue precipitate is the **insoluble** copper hydroxide.
Other transition elements react in a similar way.

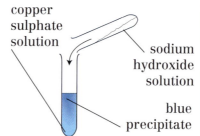

copper sulphate solution · sodium hydroxide solution · blue precipitate

copper sulphate + sodium hydroxide
$CuSO_4(aq)$ + $2NaOH(aq)$

$\downarrow$

copper hydroxide + sodium sulphate
$Cu(OH)_2(s)$ + $Na_2SO_4(aq)$

iron(II) sulphate + sodium hydroxide $\longrightarrow$ iron(II) hydroxide + sodium sulphate	**A grey-green precipitate** of iron(II) hydroxide forms.
$FeSO_4(aq)$ + $2NaOH(aq)$ $\longrightarrow$ $Fe(OH)_2(s)$ + $Na_2SO_4(aq)$	

iron(III) chloride + sodium hydroxide $\longrightarrow$ iron(III) hydroxide + sodium chloride	**An orange-brown precipitate** of iron(III) hydroxide forms.
$FeCl_3(aq)$ + $3NaOH(aq)$ $\longrightarrow$ $Fe(OH)_3(s)$ + $3NaCl(aq)$	

Precipitation reactions and reactions between acids and carbonates may be shown by essential ionic equations, e.g.:

$Cl^-(aq) + Ag^+(aq) \longrightarrow AgCl(s)$	Any solution containing chloride ions gives a white precipitate with silver ions from silver nitrate solution.
$SO_4^{2-}(aq) + Ba^{2+}(aq) \longrightarrow BaSO_4(s)$	Barium ions will give a white precipitate with the ions in any sulphate solution.
$Cu^{2+}(aq) + 2OH^-(aq) \longrightarrow Cu(OH)_2(s)$	Hydroxide ions give a blue precipitate of copper hydroxide when added to any solution containing copper(II) ions.

Questions

1. What is observed when silver nitrate solution is added to sodium chloride solution?
2. Write an equation to explain why carbon dioxide turns limewater cloudy.
3. What are silver nitrate and barium chloride used to test for?
4. Write an ionic equation for the reaction between a solution of copper ions and a hydroxide solution. What is observed during this reaction?

CALCULATIONS Formula mass and percentage composition

Chemical formulae and equations deal with **numbers of atoms**.

- The **amounts** of chemicals used in reactions may be measured by **weighing**.
- A connection between the **mass** of substance and number or **particles** present is useful.
- Each element has a **relative atomic mass** according to the relative mass of its atoms.
- Hydrogen atoms are the simplest and lightest and have an **atomic mass** of **one unit**.
- Oxygen atoms are sixteen times heavier and have an atomic mass of **sixteen units**.

H = 1 O = 16	Oxygen atoms are 16 times heavier than hydrogen atoms. 16g of oxygen atoms has the same number of atoms as 1g of hydrogen.
H = 1 Mg = 24	For samples with equal numbers of atoms, magnesium would be 24 times heavier than hydrogen.
Cu = 64 S = 32	64g of copper has the same number of atoms as 32g of sulphur. 2g of copper has same number of atoms as 1g of sulphur.

- The **formula mass** of a substance is the relative mass of one molecule or one unit as represented by its formula.
- **Formula mass** is calculated by adding the **atomic masses** according to the **numbers of atoms** in the formula.
- For water, H_2O, this means adding the masses of two hydrogen atoms and one oxygen atom.

H = 1	$H_2O = 1 + 1 + 16 = \mathbf{18}$	$FeS = 56 + 32 = \mathbf{88}$
O = 16		
Fe = 56	$SO_2 = 32 + 16 + 16 = \mathbf{64}$	$NaOH = 23 + 16 + 1 = \mathbf{40}$
S = 32		
Na = 23	$H_2SO_4 = 1 + 1 + 32 + (16 \times 4) = \mathbf{98}$	$Ca(OH)_2 = 40 + (16 + 1) + (16 + 1) = \mathbf{74}$
Ca = 40		

Percentage composition is useful in the chemical analysis of compounds.

- If a nitrogenous fertilizer is for sale, or iron ore is to be purchased, it is useful to know the **percentage** of nitrogen or iron in each sample.
- Percentage composition can be calculated from the formula of the compound concerned.
- In addition, a formula may be deduced from the percentage composition of a compound.

See page 107.

Fe = 56 $FeO = 56 + 16 = 72$
O = 16 Out of a total mass of 72, 56 is iron.   Percentage of iron = (56/72) x 100 = 77.8%.
Al = 27
N = 14 $Al_2O_3 = (27 \times 2) + (16 \times 3) = 102$
S = 32 Out of a total mass of 102, 54 is aluminium.   Percentage of aluminium = (54/102) x 100 = 52.9%.
H = 1
$(NH_4)_2SO_4 = (14 \times 2) + (1 \times 8) + 32 + (16 \times 4) = 132$
Out of a total mass of 132, 28 is nitrogen.   Percentage of nitrogen = (28/132) x 100 = 21.2%.

Questions

1. Calculate the formula mass of each of the compounds listed: MgO, $MgSO_4$, Mg_3N_2, $Mg(OH)_2$, $Mg(NO_3)_2$.
2. Calculate the percentage of copper in CuO and Cu_2O.
3. Calculate the percentage of iron in Fe_3O_4 and Fe_2O_3 and state which ore is richer in iron. Would FeO be a better ore?
4. Calculate the percentage of nitrogen in $NaNO_3$ and NH_4NO_3 and state which would be the better fertilizer.

(Mg = 24, O = 16, S = 32, N = 14, H = 1, Cu = 64, Fe = 56, Na = 23.)

CALCULATIONS Reacting masses and the mole

It is possible to calculate the mass of a substance needed to take part in a chemical reaction.

- When iron reacts with sulphur each iron atom combines with one sulphur atom.
- This means that 56g of iron should react with exactly 32g of sulphur.
- From this, the mass of sulphur which reacts with any other mass of iron can be calculated.
- The mass of iron sulphide formed in each case can also be calculated.

Fe + S $\longrightarrow$ FeS	56g of Fe react with	32g of S to give 88g of FeS
56 32 88	28g of Fe react with	16g of S to give 44g of FeS
	7g of Fe react with	4g of S to give 11g of FeS
	Xg of Fe	

Question:
Calculate the mass of sulphur needed to react with 14g of iron. (Fe = 56, S = 32)

Answer:
From the equation 56g of Fe react with 32g of S.
56g Fe $\longrightarrow$ 32g S
1g   Fe $\longrightarrow$ (32/56)g S
14g Fe $\longrightarrow$ 14 x (32/56)g S = 8 g S

Question:
Calculate the mass of iron sulphide formed when 14g of iron react with sulphur.

Answer:
From the equation 56g of Fe produce 88g of FeS.
56g Fe $\longrightarrow$ 88g FeS
1g   Fe $\longrightarrow$ (88/56)g FeS
14g Fe $\longrightarrow$ 14 x (88/56)g FeS = 22g FeS

Question:
Calculate the mass of sulphuric acid needed to react with 20g of magnesium.
(Mg = 24, H = 1, S = 32, O = 16)

Answer: Mg + H_2SO_4 $\longrightarrow$ $MgSO_4$ + H_2
                24   98   120   2
From the equation 24g of Mg react with 98g of H_2SO_4.
24g Mg $\longrightarrow$ 98g H_2SO_4
1g   Mg $\longrightarrow$ (98/24)g H_2SO_4
20g Mg $\longrightarrow$ 20 x (98/24)g H_2SO_4 = 81.7g H_2SO_4

Question:
Calculate the mass of copper oxide needed to react with sulphuric acid to produce 20g of copper sulphate.
(Cu = 64, S = 32, O = 16, H = 1)

Answer: CuO + H_2SO_4 $\longrightarrow$ $CuSO_4$ + H_2O
              80   98   160   18
From the equation 160g of $CuSO_4$ need 80g of CuO.
160g of $CuSO_4$ $\rightarrow$ 80g of CuO
1g of $CuSO_4$ $\rightarrow$ (80/160)g of CuO
20g of $CuSO_4$ $\rightarrow$ 20 x (80/160)g of CuO = 10g CuO

The atomic mass or formula mass in grams of any substance contains one **mole** of particles.

- A balance measures the mass of a substance being used in a reaction.
- It is also useful to know how many atoms or molecules are present in a reaction.
- A useful **number** of particles is 6.02 x 10^{23}, which is referred to as a **mole**.
- The atomic mass, or formula mass, will indicate how much substance is needed to provide one mole of particles.

Carbon: C = 12   12g of carbon contains 6.02 x 10^{23} atoms.
Hydrogen: H = 1   1g of hydrogen contains 6.02 x 10^{23} atoms.
Water: H_2O = 18   18g of water contains 6.02 x 10^{23} molecules.

> 6.02 x 10^{23} is known as the **Avagadro Number**. This number is also one mole.

12g of carbon is 1 mole of carbon atoms.
6g of carbon is 0.5 moles of carbon atoms.
24g of carbon is 2 moles of carbon atoms.

1 mole of carbon atoms $\longrightarrow$ 12g of carbon.
4 moles of carbon atoms $\longrightarrow$ 48g of carbon.
0.25 moles of carbon atoms $\longrightarrow$ 3g of carbon.

Questions

1. Calculate the mass of iron needed to react with 4g of sulphur and the mass of iron sulphide which would be formed.
2. *4g of copper oxide reacted completely with sulphuric acid. Calculate the mass of sulphuric acid which reacted and the mass of copper sulphate formed.*
3. *How many atoms are present in 8g of copper?*

(Fe = 56, S = 32, Cu = 64, H = 1, O = 16.)

CALCULATIONS Using moles

The formula of a compound can be calculated from the numbers of **moles** of atoms it contains.

HOW MANY MOLES ARE PRESENT IN A CERTAIN MASS OF SUBSTANCE?

Question: Calculate the number of moles in 8g of sodium hydroxide. (Na = 23, O = 16, H = 1)

Answer: sodium hydroxide, NaOH, = 40

40g NaOH $\rightarrow$ 1 mole
1g NaOH $\rightarrow$ (1/40) moles
8g NaOH $\rightarrow$ 8 x (1/40) moles = 0.2 moles

WHAT MASS IS NEEDED TO PROVIDE A CERTAIN NUMBER OF MOLES?

Question: Calculate the mass of 0.4 moles of copper sulphate. (Cu = 64, S = 32, O = 16)

Answer: copper sulphate, $CuSO_4$, = 160

1 mole $CuSO_4$ $\rightarrow$ 160g $CuSO_4$
0.4 moles $CuSO_4$ $\rightarrow$ 0.4 x 160g $CuSO_4$
$\rightarrow$ = 64g $CuSO_4$

REACTING MASSSES OF ELEMENTS CAN BE USED TO DETERMINE THE FORMULAE OF COMPOUNDS.

Question: In a reaction between iron and sulphur it was found that 7g of iron had combined with 8g of sulphur. Calculate a formula for the compound formed. (Fe = 56, S = 32)

Answer:

56g Fe $\rightarrow$ 1 mole Fe   32g S $\rightarrow$ 1 mole S
1g Fe $\rightarrow$ (1/56) moles Fe   1g S $\rightarrow$ (1/32) moles S
7g Fe $\rightarrow$ 7 x (1/56) moles = 0.125 moles of Fe   8g S $\rightarrow$ 8 x (1/32) moles = 0.25 moles of S

This gives a formula '$Fe_{0.125}S_{0.25}$'; to obtain a whole number formula it is necessary to divide each number by the smaller one, 0.125, giving FeS_2.

Moles of Fe = (0.125/0.125) = 1
Moles of S = (0.25/0.125) = 2

The formula becomes **FeS_2**. This is the simplest formula (**empirical formula**) from these data. The **true** formula could be FeS_2 or any multiple of this such as Fe_2S_4 or Fe_3S_6, but more data is needed to confirm which.

ANALYSIS OF COMPOUNDS MAY INDICATE THE PERCENTAGES OF ELEMENTS PRESENT. FORMULAE OF COMPOUNDS CAN BE DETERMINED FROM PERCENTAGE COMPOSITION.

Question: A hydrocarbon was found to contain 14.3% hydrogen and 85.7% carbon by mass. Calculate its formula. (C = 12, H = 1)

Answer: In 100g of the compound there would be 14.3g of hydrogen and 85.7g of carbon.

12g C $\rightarrow$ 1 mole C   1g H $\rightarrow$ 1 mole H atoms
1g C $\rightarrow$ (1/12) moles C   14.3g H $\rightarrow$ 14.3 moles H
85.7g C $\rightarrow$ 85.7 x (1/12) moles C = 7.15 moles of C

This gives a formula '$C_{7.15}H_{14.3}$'; to obtain a whole number formula it is necessary to divide each number by the smaller one, 7.15, giving CH_2.

Moles of C atoms = (7.15/7.15) = 1
Moles of H atoms = (14.3/7.15) = 2

The formula becomes **CH_2**, but this is the simplest formula (**empirical formula**) and the **true** formula could be CH_2, C_2H_4, C_3H_6 and so on. More information is needed to discover the true (molecular) formula.

MOLES CAN BE USED TO CALCULATE THE VOLUME OF GAS PRODUCED IN A REACTION.

- There is a direct connection between the volume of a gas and the number of moles it contains.
- At 25°C and 1 atm pressure one mole of gas occupies 24 000cm^3 (24dm^3).

Question: What volume of hydrogen is produced when 6g of magnesium react with sulphuric acid?

Answer:

$$Mg + H_2SO_4 \rightarrow MgSO_4 + H_2$$
$$24 \qquad 98 \qquad\quad 120 \qquad 2$$

| H_2 = 2, so 2g of hydrogen is one mole of gas. |

24g Mg $\rightarrow$ 1 mole H_2; so 1g Mg $\rightarrow$ (1/24) moles H_2; and 6g Mg $\rightarrow$ 6 x (1/24) moles H_2 = 0.25 moles H_2.

1 mole H_2 = 24dm^3; so 0.25 moles = 0.25 x 24dm^3 = 6dm^3 of H_2.

Questions

1. Calculate the number of moles in (a) 5g NaOH, (b) 5g of $CuSO_4$.
2. Calculate the mass of (a) 3 moles of NaOH, (b) 3 moles of $CuSO_4$.
3. 62.1g of lead combined with 6.4g of oxygen. Calculate the formula of the oxide of lead produced.

Na = 23	O = 16
Cu = 64	H = 1
Pb = 207	S = 32

RATES OF REACTION
Slow and fast reactions
Measuring rates

Rate measures change in a single unit of time – a second, a minute, an hour, a week ………
There is great variation in the rates of chemical reactions.

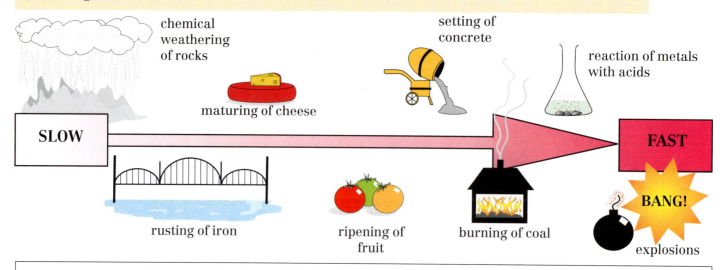

chemical weathering of rocks

setting of concrete

reaction of metals with acids

maturing of cheese

SLOW

FAST

BANG!

rusting of iron

ripening of fruit

burning of coal

explosions

Experiment   The reaction between marble chips (calcium carbonate) and dilute hydrochloric acid

When marble chips (calcium carbonate) are added to dilute hydrochloric acid:
- The mixture fizzes as a colourless gas is given off – this gas is carbon dioxide.
- The marble chips slowly disappear.
- The reaction stops when either all the acid or all the marble chips are used up.

REACTANTS ⟶ PRODUCTS

calcium carbonate + hydrochloric acid ⟶ calcium chloride + water + carbon dioxide

$$CaCO_3\,(s) \; + \; 2HCl\,(aq) \longrightarrow CaCl_2\,(aq) \; + \; H_2O\,(l) \; + \; CO_2\,(g)$$

The progress of a reaction can be followed:
(a) by measuring the rate of disappearance of a reactant or
(b) by measuring the rate of appearance of a product.

UNDERSTANDING GRAPHS SHOWING RATES OF REACTION

Example: Appearance of a product

mass/g or volume/cm³ of product (y-axis)
time/minutes (x-axis)

1. The reaction rate is fastest at the beginning – the curve is steepest.
2. The reaction rate is slowing down – the curve is less steep.
3. The reaction has stopped – the curve is flat.

Questions

1. For each of the reactions mentioned at the top of the page, give a suitable time over which it will take place.
2. What would you measure to find the rate of the reaction between calcium carbonate and dilute hydrochloric acid?
3. Why does the reaction stop eventually?
4. *Why is the reaction rate fastest at the beginning?*
5. *Why does the reaction slow down as it proceeds?*
6. *How could you test the gas formed to show that it is carbon dioxide?*
7. *Sketch the graph you would expect to obtain if you measured the disappearance of a reactant.*
8. *Which of the following curves shows a faster reaction rate? Explain your answer.*

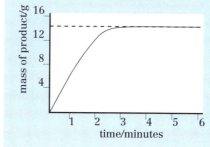

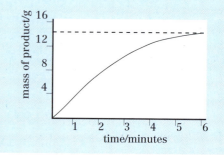

RATES OF REACTION   The effect of temperature

A reaction goes **faster** when the temperature is **raised**.
An increase in temperature of **10C°** approximately **doubles** the rate.

Experiment   **Zinc and dilute sulphuric acid**

Experiment A

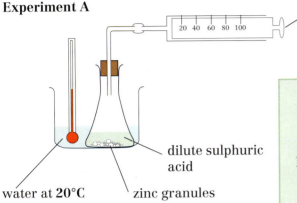

gas syringe – the plunger moves outwards as gas is given off

dilute sulphuric acid

water at **20°C**   zinc granules

Add the metal and replace the bung immediately. Start timing as soon as the plunger in the syringe starts to move.

The volume of gas is measured every minute.

When zinc granules are added to dilute sulphuric acid:
- The mixture fizzes as a colourless gas is given off – this gas is hydrogen.
- The zinc granules slowly disappear.
- The reaction stops when either all the acid or all the zinc is used up.

zinc + sulphuric acid $\longrightarrow$ zinc sulphate + hydrogen

$Zn\ (s)\ +\quad H_2SO_4\ (aq)\quad \longrightarrow\quad ZnSO_4\ (aq)\quad +\quad H_2\ (g)$

Experiment B

The experiment is repeated using water at **30°C**.
To allow a **fair comparison** to be made, only the temperature at which the reaction takes place must be changed.
Other conditions must be kept constant:
- **mass** of zinc granules
- **sizes** of zinc granules
- **concentration** of acid
- **volume** of acid.

This graph shows the results of such an experiment.

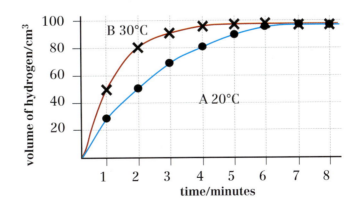

RATES OF REACTION   The effect of concentration and pressure

A reaction involving **solutions** goes **faster** when the **concentration** of a reactant is **increased**.
A reaction involving **gases** goes **faster** when the **pressure** is **increased**.

Experiment   Sodium thiosulphate solution and dilute hydrochloric acid

When sodium thiosulphate solution is added to dilute hydrochloric acid, the mixture becomes cloudy because a fine yellow precipitate is produced. This precipitate is sulphur.

A cross, drawn on a piece of paper, is viewed through the solution from above. As the reaction proceeds, the cross gradually disappears. The faster the reaction, the more quickly the cross disappears. The time it takes for the sulphur to obscure the cross is an indication of the rate of the reaction.

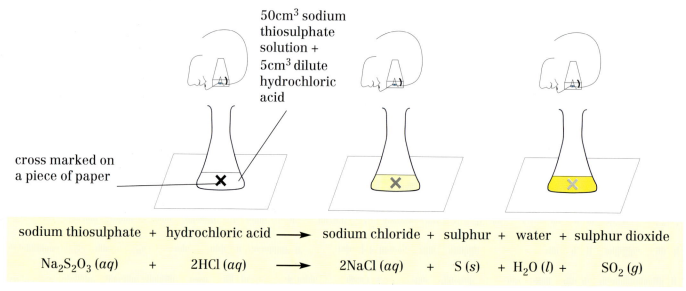

50cm³ sodium thiosulphate solution + 5cm³ dilute hydrochloric acid

cross marked on a piece of paper

sodium thiosulphate	+	hydrochloric acid	$\longrightarrow$	sodium chloride	+	sulphur	+	water	+	sulphur dioxide
$Na_2S_2O_3 (aq)$	+	$2HCl (aq)$	$\longrightarrow$	$2NaCl (aq)$	+	$S (s)$	+	$H_2O (l)$ +		$SO_2 (g)$

The effect of concentration on the rate of the reaction can be investigated by repeating the experiment using different concentrations of sodium thiosulphate solution. Different concentrations can be obtained by diluting a standard solution of sodium thiosulphate. The total volume used each time is 50cm³. At the start of the experiment, 5cm³ dilute hydrochloric acid is added.

In an experiment carried out as above, the following results were obtained.

volume of sodium thiosulphate solution/cm³	volume of water/cm³	volume of hydrochloric acid/cm³	time taken for cross to disappear/seconds
50	0	5	65
45	5	5	68
40	10	5	75
35	15	5	90
30	20	5	110
25	25	5	135
20	30	5	180
15	35	5	300
10	40	5	540

GASES

In a reaction involving gases, increasing the pressure produces an effect similar to increasing the concentration of reactants in solution.

Increasing the pressure pushes reacting molecules closer together.

Questions

1. Using the results in the table plot a graph of time taken for the cross to disappear against volume of sodium thiosulphate solution. Draw in a curve of 'best fit'.
2. What can you conclude from these results?
3. *What features of the experiment allow a fair comparison to be made of the rates of the reaction at different concentrations of sodium thiosulphate solution?*
4. *If two students were working together on this experiment, why would it be important for the same person to view the cross each time?*
5. *If the concentration of sodium thiosulphate solution were kept constant each time what would be the effect of varying the concentration of hydrochloric acid?*
6. *Using sodium thiosulphate solution and dilute hydrochloric acid describe how you could carry out an experiment to investigate the effect of temperature on the reaction.*

RATES OF REACTION   The effect of surface area

The rate of a reaction **increases** when the **surface area** of a solid reactant is **increased**.
For the same mass of reactant a number of smaller pieces have a greater surface area than one large piece.

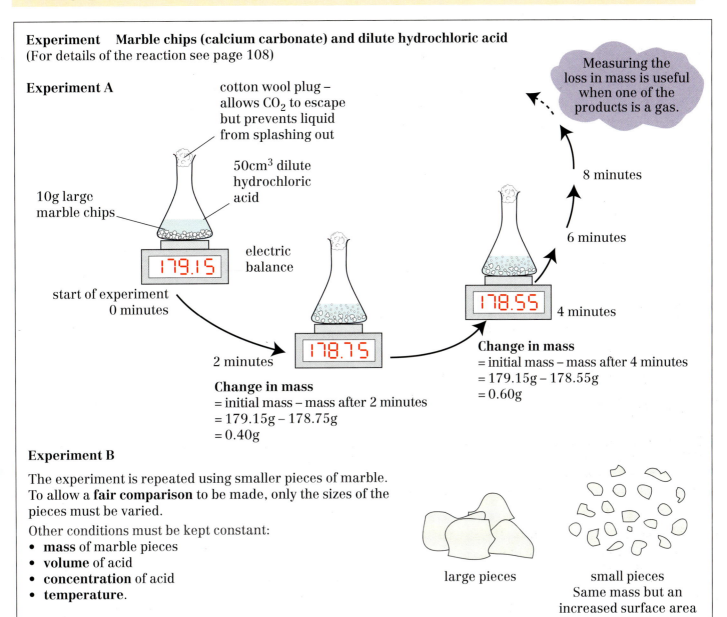

Experiment   Marble chips (calcium carbonate) and dilute hydrochloric acid
(For details of the reaction see page 108)

Experiment A

cotton wool plug –
allows CO_2 to escape
but prevents liquid
from splashing out

$50cm^3$ dilute
hydrochloric
acid

10g large
marble chips

electric
balance

179.15

start of experiment
0 minutes

2 minutes

178.75

Change in mass
= initial mass – mass after 2 minutes
= 179.15g – 178.75g
= 0.40g

178.55

4 minutes

6 minutes

8 minutes

Measuring the
loss in mass is useful
when one of the
products is a gas.

Change in mass
= initial mass – mass after 4 minutes
= 179.15g – 178.55g
= 0.60g

Experiment B

The experiment is repeated using smaller pieces of marble.
To allow a **fair comparison** to be made, only the sizes of the
pieces must be varied.

Other conditions must be kept constant:
- **mass** of marble pieces
- **volume** of acid
- **concentration** of acid
- **temperature**.

large pieces

small pieces
Same mass but an
increased surface area

In an experiment carried out as above, the following results were obtained.

Time (minutes)	0	2	4	6	8	10	12	14	16	18	20
Loss in mass A (large pieces) (g)	0.00	0.40	0.60	0.75	0.83	0.90	1.00	1.05	1.10	1.10	1.10
Loss in mass B (small pieces) (g)	0.00	0.60	0.80	0.90	1.00	1.08	1.10	1.10	1.10	1.10	1.10

Questions

1. Using the results in the table plot graphs of loss in mass (vertical axis) against time (horizontal axis) for large and small pieces. Plot both sets of results on the same axes and draw in curves of 'best fit'.
2. Why is there a loss in mass as the experiment proceeds?
3. Which experiment, A or B, proceeds at a faster rate?
4. Why do the reactions proceed at different rates?
5. Why do both experiments produce the same total loss in mass (1.10g)?
6. Give four factors which are controlled in these experiments in order to make them a fair test.

RATES OF REACTION   The effect of a catalyst

A catalyst is a substance which **speeds up** the rate of a chemical reaction **without being chemically changed** itself.
A catalyst is not used up – it can be used again.
A small amount of a catalyst can bring about a large amount of chemical change.

The decomposition of hydrogen peroxide

Hydrogen peroxide is a colourless liquid, which **decomposes** very slowly to give water and oxygen.
If a small amount of manganese(IV) oxide is added, the mixture fizzes vigorously as a colourless gas is given off. This gas is oxygen.

hydrogen peroxide $\longrightarrow$ water + oxygen

$$2H_2O_2 \, (aq) \longrightarrow 2H_2O \, (l) + O_2 \, (g)$$

The test for oxygen
A glowing splint is placed in the gas.
The splint relights.

Hydrogen peroxide is the only reactant.

This reaction is an example of **decomposition**.

Over 90% of industrial processes use catalysts

Examples	Process		Product	Catalyst
	Haber Process	$\longrightarrow$	ammonia	iron
	Contact Process	$\longrightarrow$	sulphuric acid	vanadium(V) oxide
	Fermentation	$\longrightarrow$	ethanol (brewing, wine making)	enzymes in yeast
	Hydrogenation	$\longrightarrow$	margarine	nickel

The action of a catalyst

The surface of a catalyst is important.
When a reactant comes into contact with the catalyst, chemical bonds are broken more easily and the reaction is speeded up. If a catalyst is divided, the surface area is increased, the frequency of collisions increases, and the rate of the reaction increases.

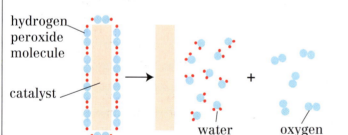
hydrogen peroxide molecule
catalyst
water molecule
oxygen molecule

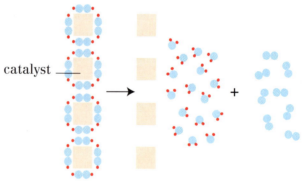
catalyst

How a catalyst might help this reaction

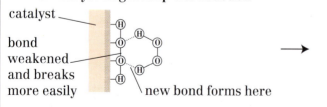
catalyst
bond weakened and breaks more easily
new bond forms here

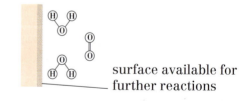
surface available for further reactions

Questions

1. What is a catalyst?
2. In the decomposition of hydrogen peroxide, name:
   (a) the reactant
   (b) the catalyst for the reaction
   (c) the products.
3. Describe the test for oxygen.
4. *Draw a diagram of the apparatus which could be used to measure the volume of oxygen produced.*
5. *If 1 g of manganese (IV) oxide is added to 50 cm3 of hydrogen peroxide solution, how could you show that the catalyst had not been used up?*
6. *Why might impurities in the reaction mixture slow down the rate of a reaction which uses a catalyst?*
7. *What is meant by a decomposition reaction?*
8. *Suggest a reason why catalysts might be important in industry.*

RATES OF REACTION   Collision theory

For a reaction to take place, particles must **collide** with each other.
Not all collisions result in a reaction – the particles must have **sufficient energy**.

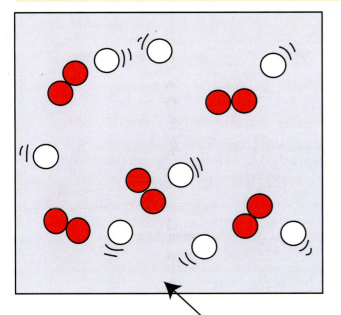

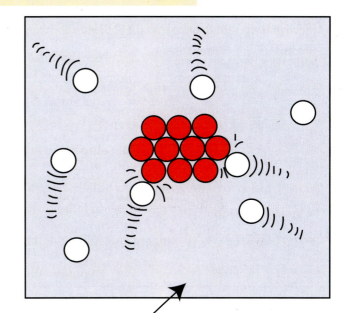

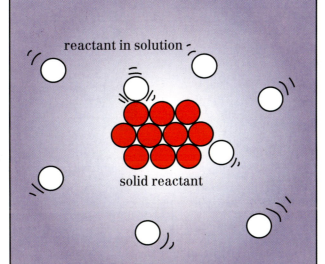

reactant in solution

solid reactant

Increasing the surface area of a **solid reactant** increases the rate of a reaction.

• There are **more collisions** in a given time.

Increasing the pressure in reactions involving **gases** pushes particles closer together.

• There are **more collisions** in a given time.

Increasing the concentration of a **reactant in solution** increases the rate of the reaction.

• There are **more collisions** in a given time.

Increasing the temperature increases the rate of a reaction.

• The particles are moving faster – there are **more collisions** in a given time.

• The particles have **more energy** – there are **more successful collisions** in a given time.

Using a catalyst increases the rate of the reaction because it lowers the energy needed for a successful collision – the **activation energy**.
The lower the activation energy, the **more successful collisions** there are.

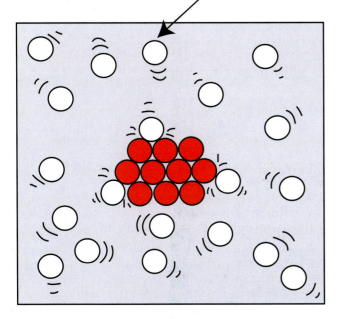

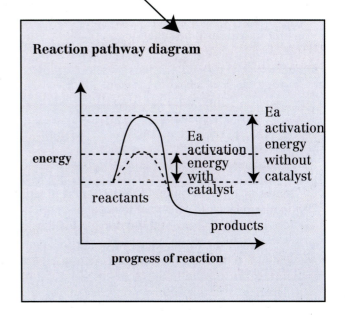

Reaction pathway diagram

energy

Ea activation energy without catalyst

Ea activation energy with catalyst

reactants

products

progress of reaction

ENZYMES Fermentation

Living cells use chemical reactions to produce new materials.
Enzymes are biological catalysts – catalysts produced by living organisms.
Enzymes are proteins.

Experiment   Fermentation

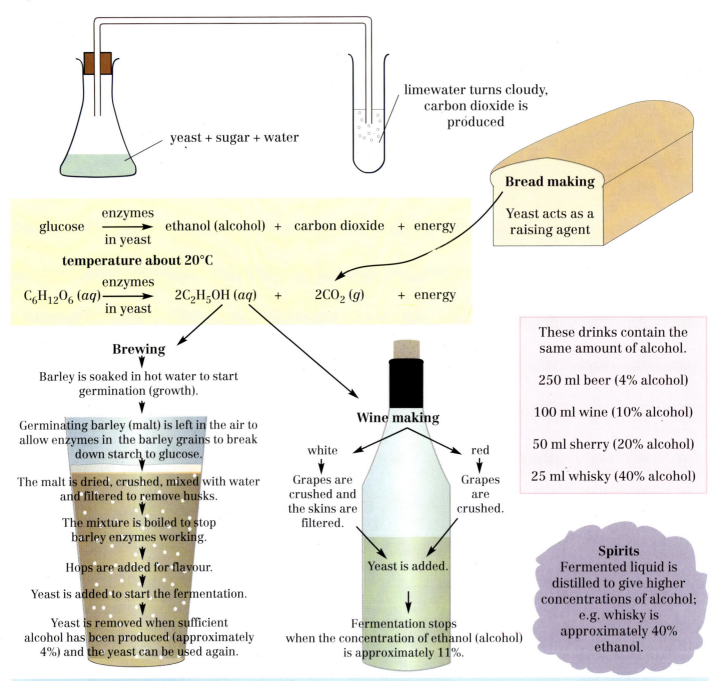

limewater turns cloudy, carbon dioxide is produced

yeast + sugar + water

Bread making

Yeast acts as a raising agent

$$\text{glucose} \xrightarrow[\text{in yeast}]{\text{enzymes}} \text{ethanol (alcohol)} + \text{carbon dioxide} + \text{energy}$$

temperature about 20°C

$$C_6H_{12}O_6 \ (aq) \xrightarrow[\text{in yeast}]{\text{enzymes}} 2C_2H_5OH \ (aq) + 2CO_2 \ (g) + \text{energy}$$

Brewing

Barley is soaked in hot water to start germination (growth).

Germinating barley (malt) is left in the air to allow enzymes in the barley grains to break down starch to glucose.

The malt is dried, crushed, mixed with water and filtered to remove husks.

The mixture is boiled to stop barley enzymes working.

Hops are added for flavour.

Yeast is added to start the fermentation.

Yeast is removed when sufficient alcohol has been produced (approximately 4%) and the yeast can be used again.

Wine making

white — Grapes are crushed and the skins are filtered.

red — Grapes are crushed.

Yeast is added.

Fermentation stops when the concentration of ethanol (alcohol) is approximately 11%.

These drinks contain the same amount of alcohol.

250 ml beer (4% alcohol)

100 ml wine (10% alcohol)

50 ml sherry (20% alcohol)

25 ml whisky (40% alcohol)

Spirits
Fermented liquid is distilled to give higher concentrations of alcohol; e.g. whisky is approximately 40% ethanol.

Questions
1. Why are enzymes called 'biological catalysts'?
2. Which product of the fermentation process enables yeast to be used as a raising agent in bread making?
3. Explain why each of the following stages is carried out in the brewing process:
    (a) soaking of barley grains in water
    (b) leaving germinating barley grains in the air
    (c) filtering of crushed barley grains once water has been added
    (d) boiling the filtered liquid
    (e) adding hops
    (f) adding yeast.
4. What is the main difference between the fermentation processes for making white and red wine?
5. Why does whisky have a higher concentration of alcohol than wine?
6. What is the approximate alcohol concentration (expressed as % by volume) of
    (a) beer   (b) wine   (c) whisky?
7. *What is meant by the term 'anaerobic'?*
8. *What is the energy released in fermentation used for by the yeast cells?*

ENZYMES Temperature, pH and enzyme action

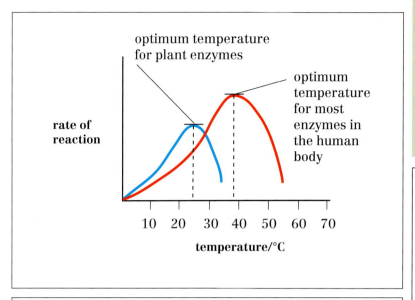

optimum temperature for plant enzymes

optimum temperature for most enzymes in the human body

rate of reaction

temperature/°C

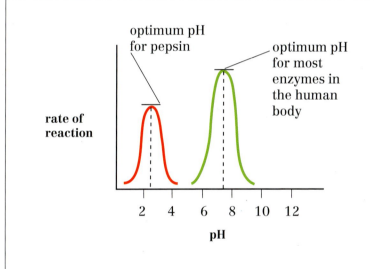

optimum pH for pepsin

optimum pH for most enzymes in the human body

rate of reaction

pH

Questions

1. Why are enzymes described as being specific?
2. What is the 'active site' of an enzyme?
3. Why does enzyme activity decrease above the optimum temperature?
4. What is the optimum temperature for pepsin?
5. What is the optimum pH for pepsin?
6. How do enzymes speed up the rate of a reaction?
7. *Why is it important that our body temperature should remain constant?*
8. *Pepsin is an enzyme which catalyses the digestion (breakdown) of proteins. Pepsin is contained in gastric juice, made in the stomach. Why does gastric juice also contain hydrochloric acid?*
9. *If enzymes in the small intestine have an optimum pH of 8, what will need to happen as food passes from the stomach to the small intestine?*
10. *What would you expect the optimum temperature to be for enzymes in bacteria found in hot springs?*
11. *Why would most plant enzymes not have an optimum temperature higher than about 25°C?*
12. *In brewing, a temperature of 18–20°C is used for the fermentation stage.*
    (a) Why is a lower temperature not used?
    (b) Why is a higher temperature not used?

Enzymes speed up biological reactions, but they are **unlike other catalysts** because:
- They are **specific** in the reactions they catalyse – they catalyse only one reaction or one type of reaction, e.g. fermentation, digestion.
- They work best at an **optimum temperature** – above this temperature they lose their shape and become **denatured**.
- They work best at an **optimum pH** – the degree of acidity or alkalinity.

The shape of an enzyme is important

This model of enzyme action shows why enzymes are specific

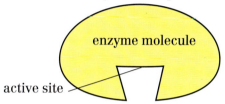

enzyme molecule

active site

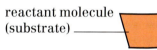

reactant molecule (substrate)

The reactant molecule (substrate) fits into the 3-dimensional shape (active site) of the enzyme.

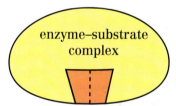

enzyme–substrate complex

The complex formed helps the reactant to break down more easily.

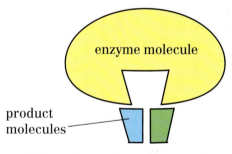

enzyme molecule

product molecules

When the product molecules have been released the enzyme can be used again.

NOTE

In a reaction involving more than one reactant, the complex formed with the enzyme can bring reacting molecules together so that they react more easily.

Enzymes function by lowering the activation energy so that the reaction takes place at a lower temperature.

Heating an enzyme above its optimum temperature changes the 3-dimensional shape so that the substrate no longer fits the active site.

ENZYMES Uses of enzymes and biotechnology

Enzymes enable many reactions, which would otherwise require energy-demanding equipment, to take place at normal temperatures and pressures.
Energy and equipment costs are important considerations in industrial processes.
Enzymes are also used in the home and in medicine.

Yoghurt-making

Milk is heat-treated to 90°C for 15–30 minutes to kill bacteria and, after cooling, a culture of *Lactobacillus* (a bacterium) is added. This ferments the sugar (lactose) in milk to lactic acid, which causes milk solids to separate.

BIOTECHNOLOGY

is the commercial or industrial use of micro-organisms, e.g. yeast and bacteria or substances from them such as enzymes and genes.

Cheese-making

Milk is heat-treated to 90°C for 15–30 minutes to kill bacteria and, after cooling, a culture of *Streptococcus* (a bacterium) is added. This separates the milk into curds (solid) and whey (liquid). The curds are pressed and dried to make cheese.

Biological detergents

Clothes may be stained by fat, e.g. butter and margarine and by proteins, e.g. blood and egg yolk. Biological detergents contain **proteases** (enzymes which digest proteins) and **lipases** (enzymes which digest fats). These enzymes have an optimum temperature of approximately 40°C.

Production of glucose syrup

Cheap starch from e.g. corn or maize is digested by **carbohydrases** (enzymes which break down carbohydrates) to give glucose syrup.

USES OF ENZYMES

In medicine

Blood clots (thromboses) can be broken down by **proteases** (enzymes which digest proteins). **Enzymes** in Clinistix™ are used to detect glucose in urine.

Production of fruit juices

Enzymes may be used to release juice from cells in fruit.

Slimming foods

An enzyme called **isomerase** converts glucose to fructose, a sweeter sugar which can be used in smaller quantities.

Baby foods

Proteins in baby food may be pre-digested by **proteases** (enzymes which digest proteins).

Soft-centred chocolates

An enzyme called **invertase** is used to soften the centres of some chocolates.

Industrial processes usually

- stabilise the enzyme so that it works for a long time
- trap or immobilise enzymes in an inert solid support or carrier such as alginate beads – this allows liquids to move past the enzyme without the enzyme being washed away
- allow continuous production rather than batch process.

Questions

1. What do you understand by the term biotechnology?
2. Why do many industries use enzymes as part of a production process?
3. Describe one use of an enzyme in medicine.
4. Describe one use of an enzyme in food production.
5. What reactions are carried out by each of the following enzymes:
   (a) carbohydrases   (c) lipases
   (b) proteases   (d) isomerase?
6. *Apart from improved performance, what is another advantage of using biological detergents?*
7. *Give one disadvantage of using biological detergents.*

REVERSIBLE REACTIONS Examples of reversible reactions

In some chemical reactions the products can react to give the original reactants.

forward reaction

A + B $\rightleftharpoons$ C + D

back reaction

This symbol shows that a reaction is reversible

Experiment   Heating copper sulphate crystals

blue crystals – hydrated copper sulphate

white powder – anhydrous copper sulphate

water

White powder turns blue as water is added to anhydrous copper sulphate.

This reaction is **endothermic**

This reaction is **exothermic** – the heat of the reaction causes some of the water to turn to steam.

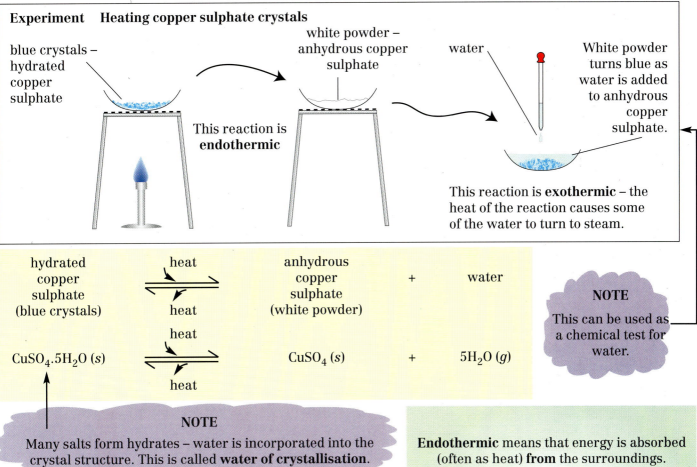

hydrated copper sulphate (blue crystals)	heat $\rightleftharpoons$ heat	anhydrous copper sulphate (white powder)	+	water
$CuSO_4.5H_2O$ (s)	heat $\rightleftharpoons$ heat	$CuSO_4$ (s)	+	$5H_2O$ (g)

NOTE
This can be used as a chemical test for water.

NOTE
Many salts form hydrates – water is incorporated into the crystal structure. This is called **water of crystallisation**.

Endothermic means that energy is absorbed (often as heat) **from** the surroundings.

Exothermic means that energy is transferred (often as heat) **to** the surroundings.

If the forward reaction is endothermic, the back reaction is exothermic – the amount of energy transferred is the same in each case.

Experiment   Heating ammonium chloride crystals

white crystals of ammonium chloride

A white solid forms on the cooler parts of the test tube; this is ammonium chloride.

The crystals disappear from the bottom of the test tube.

ammonium chloride (white crystals) $\rightleftharpoons$ ammonia + hydrogen chloride (colourless gases)

NH_4Cl (s) $\rightleftharpoons$ NH_3 (g) + HCl (g)

Ammonia is the only common alkaline gas – it turns moist red litmus paper blue.

Questions
1. Explain the term 'reversible reaction'.
2. How would you use a sample of anhydrous copper sulphate to show that a liquid contained water?
3. How would you test a colourless gas to confirm that it was ammonia?
4. What is 'water of crystallisation'?
5. What symbol is used in an equation to show that a reaction is reversible?
6. *Describe a physical test to confirm whether a liquid is pure water.*
7. *Give two examples each of:*
   (a) an endothermic reaction
   (b) an exothermic reaction
   other than those shown on this page.

REVERSIBLE REACTIONS The Haber Process

Many industrial processes use reversible reactions as a whole or a part of the process.
In a closed system, **equilibrium** will be reached – the **forward** and **back reactions** occur at the **same rate** and the reaction will not go to completion.
The relative amounts of all reacting substances depend on the conditions of the reaction.

The Haber Process – The industrial production of ammonia

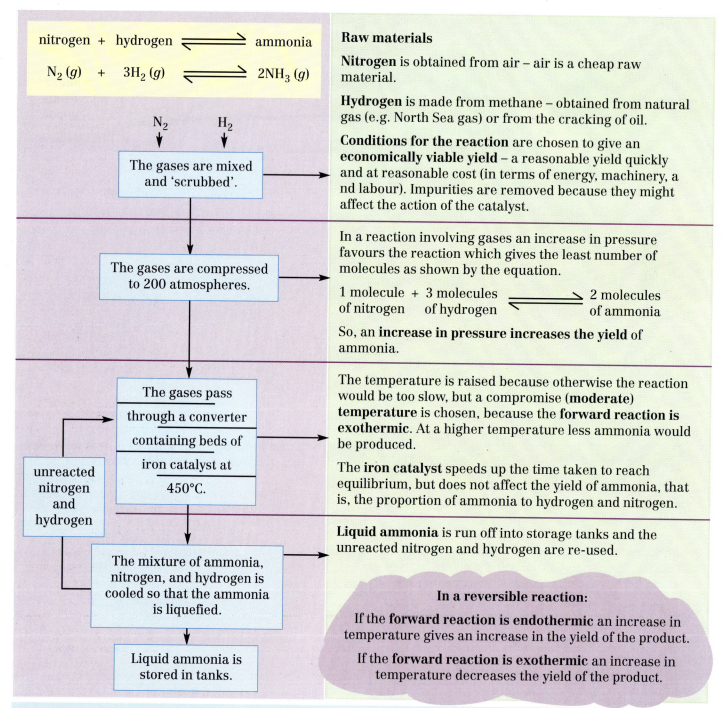

nitrogen + hydrogen ⇌ ammonia

$$N_2\,(g) + 3H_2\,(g) \rightleftharpoons 2NH_3\,(g)$$

N_2 H_2

The gases are mixed and 'scrubbed'.

The gases are compressed to 200 atmospheres.

The gases pass through a converter containing beds of iron catalyst at 450°C.

unreacted nitrogen and hydrogen

The mixture of ammonia, nitrogen, and hydrogen is cooled so that the ammonia is liquefied.

Liquid ammonia is stored in tanks.

Raw materials

Nitrogen is obtained from air – air is a cheap raw material.

Hydrogen is made from methane – obtained from natural gas (e.g. North Sea gas) or from the cracking of oil.

Conditions for the reaction are chosen to give an **economically viable yield** – a reasonable yield quickly and at reasonable cost (in terms of energy, machinery, and labour). Impurities are removed because they might affect the action of the catalyst.

In a reaction involving gases an increase in pressure favours the reaction which gives the least number of molecules as shown by the equation.

1 molecule + 3 molecules ⇌ 2 molecules
of nitrogen   of hydrogen   of ammonia

So, an **increase in pressure increases the yield** of ammonia.

The temperature is raised because otherwise the reaction would be too slow, but a compromise (**moderate**) **temperature** is chosen, because the **forward reaction is exothermic**. At a higher temperature less ammonia would be produced.

The **iron catalyst** speeds up the time taken to reach equilibrium, but does not affect the yield of ammonia, that is, the proportion of ammonia to hydrogen and nitrogen.

Liquid ammonia is run off into storage tanks and the unreacted nitrogen and hydrogen are re-used.

In a reversible reaction:

If the **forward reaction is endothermic** an increase in temperature gives an increase in the yield of the product.

If the **forward reaction is exothermic** an increase in temperature decreases the yield of the product.

Questions
1. Explain the term 'economically viable yield'.
2. From where are the raw materials for the Haber Process obtained?
3. Why are the reacting gases scrubbed as they are mixed?
4. What temperature and pressure are used in the Haber Process?
5. What is the effect of the iron catalyst?
6. What happens to unreacted nitrogen and hydrogen?
7. *If increasing the pressure increases the yield of ammonia, why do you think a pressure greater than 200 atmospheres is not used?*
8. *What would happen to the yield of ammonia if:*
   (a) a higher temperature than 450°C were used
   (b) a lower temperature than 450°C were used?
9. *Why is the mixture of nitrogen and hydrogen passed over several beds of iron catalyst?*
10. *Why does a mixture of ammonia, nitrogen, and hydrogen leave the converter?*

REVERSIBLE REACTIONS Uses of ammonia and nitric acid

Much of the ammonia produced in the Haber Process is used for the manufacture of nitric acid. The most important use of nitric acid is the manufacture of artificial fertilizers. Fertilizers are important in agriculture, but cause problems when overused.

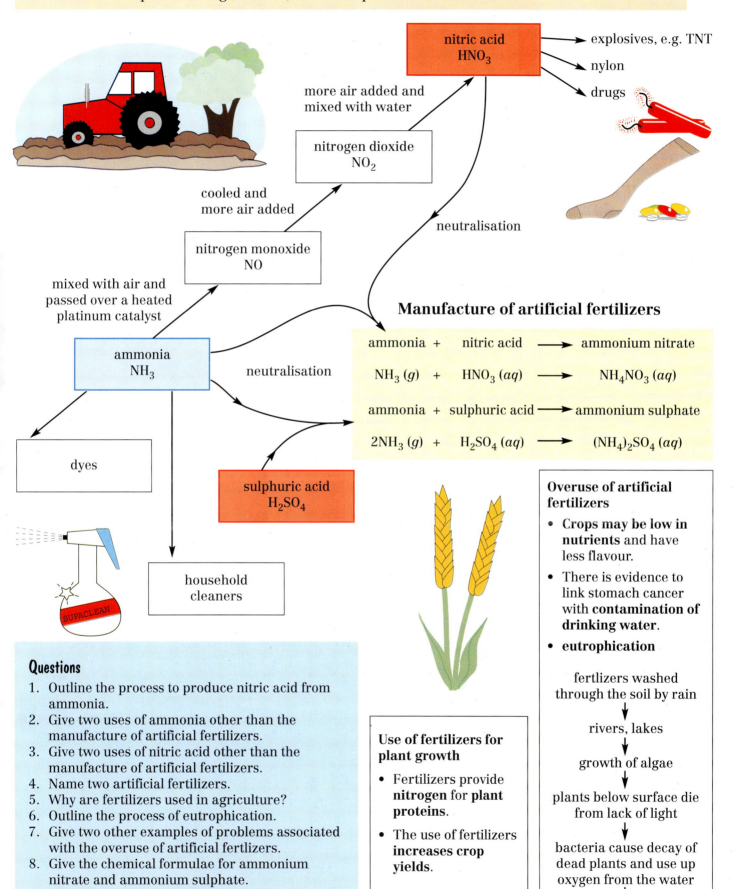

nitric acid HNO_3 → explosives, e.g. TNT

→ nylon

→ drugs

nitrogen dioxide NO_2

more air added and mixed with water

cooled and more air added

nitrogen monoxide NO

neutralisation

mixed with air and passed over a heated platinum catalyst

ammonia NH_3

neutralisation

dyes

household cleaners

sulphuric acid H_2SO_4

Manufacture of artificial fertilizers

ammonia + nitric acid ⟶ ammonium nitrate

$$NH_3 \, (g) + HNO_3 \, (aq) \longrightarrow NH_4NO_3 \, (aq)$$

ammonia + sulphuric acid ⟶ ammonium sulphate

$$2NH_3 \, (g) + H_2SO_4 \, (aq) \longrightarrow (NH_4)_2SO_4 \, (aq)$$

Use of fertilizers for plant growth

- Fertilizers provide **nitrogen** for **plant proteins**.
- The use of fertilizers **increases crop yields**.

Overuse of artificial fertilizers

- **Crops may be low in nutrients** and have less flavour.
- There is evidence to link stomach cancer with **contamination of drinking water**.
- **eutrophication**

fertilizers washed through the soil by rain

↓

rivers, lakes

↓

growth of algae

↓

plants below surface die from lack of light

↓

bacteria cause decay of dead plants and use up oxygen from the water

↓

death of fish and other animal life

Questions

1. Outline the process to produce nitric acid from ammonia.
2. Give two uses of ammonia other than the manufacture of artificial fertilizers.
3. Give two uses of nitric acid other than the manufacture of artificial fertilizers.
4. Name two artificial fertilizers.
5. Why are fertilizers used in agriculture?
6. Outline the process of eutrophication.
7. Give two other examples of problems associated with the overuse of artificial fertlizers.
8. Give the chemical formulae for ammonium nitrate and ammonium sulphate.
9. *What do you think is meant by a natural fertilizer?*
10. *What is meant by the term 'organic farming'?*

ENERGY CHANGES   Energy changes

The energy involved in chemical reactions concerns breaking and making chemical bonds.

- Reactions which give out energy as heat are **exothermic**.
- Reactions which take in heat energy are **endothermic**.

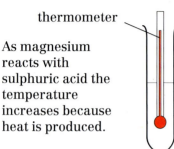

thermometer

As magnesium reacts with sulphuric acid the temperature increases because heat is produced.

Heat has to be supplied to make magnesium carbonate react and turn into magnesium oxide.

Heat is needed for magnesium to start burning but then it will continue to react and give out heat.

bright white flame

- Energy changes occurring during reactions can be shown on **reaction pathway diagrams**.
- ΔH shows the overall **energy change** and E_a shows the **activation energy**.
- Activation energy is the energy needed to start the reaction by **breaking** existing **bonds**.

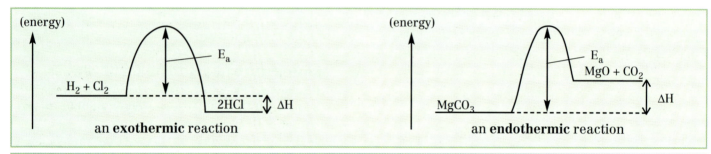

(energy)

E_a

$H_2 + Cl_2$

2HCl $\updownarrow$ ΔH

an **exothermic** reaction

(energy)

E_a
$MgO + CO_2$

$MgCO_3$

ΔH

an **endothermic** reaction

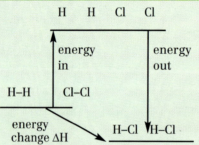

Before hydrogen and chlorine atoms can combine, the H_2 and Cl_2 molecules have to be broken apart, but this requires energy to **break existing bonds**.

H   H   Cl   Cl

energy in

energy out

H–H   Cl–Cl

energy change ΔH

H–Cl   H–Cl

Once the hydrogen and chlorine atoms are free they can combine with each other, and when new bonds are formed **energy is given out**.

The overall change is **exothermic** because more energy is given out than is taken in.

$$H_2(g) + Cl_2(g) \longrightarrow 2HCl(g)$$

Bond energy is the amount of energy needed to break a **covalent** bond. It also indicates the **energy** given out when a **bond forms**. The difference between the energy needed to break bonds and the energy given out when new bonds form represents the overall energy change, ΔH, for the reaction.

EXAMPLE: Calculate the energy change for the reaction between chlorine and hydrogen.
    Bond energies (kJ/mol) Cl–Cl = 243, H–H = 436, H–Cl = 432

Bonds broken	= 1 x Cl–Cl and 1 x H–H	= 243 + 436	= 679 kJ supplied.
Bonds formed	= 2 x H–Cl	= 2 x 432	= 864 kJ given out.
Overall difference	= 864 - 679		= **185 kJ given out.**

The reaction is therefore **exothermic** by 185 kJ per mole and $\Delta H = -185$ kJ/mol.

Note: The energy change ΔH is conventionally given a **minus** value for **exothermic** reactions because the products of the reaction have **less energy** than the starting materials; ΔH for an **endothermic** reaction is given a **positive** sign because the reaction products have **more energy** than the starting materials.

Questions

1. *Draw a reaction pathway diagram for hydrogen burning in oxygen to make water.*
2. *Calculate the energy change for the reaction between nitrogen and hydrogen.*

Bond energies (kJ/mole): H–H = 436; N–H = 391; N–N = 945

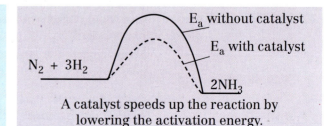

E_a without catalyst

E_a with catalyst

$N_2 + 3H_2$

$2NH_3$

A catalyst speeds up the reaction by lowering the activation energy.

REVIEW QUESTIONS   Patterns of behaviour III

1. For the substances listed below state what type of reaction would occur and write an equation for each one:
   (a) When magnesium reacts with chlorine
       $Mg(s)$ + $Cl_2(g)$
   (b) When carbon monoxide reacts with oxygen
       $CO(g)$ + $O_2(g)$
   (c) When potassium hydroxide reacts with hydrochloric acid
       $KOH(aq)$ + $HCl(aq)$
   (d) When methane burns in oxygen
       $CH_4(g)$ + $O_2(g)$
   (e) When lead oxide is heated with carbon
       $PbO(s)$ + $C(s)$
   (f) When iron reacts with copper sulphate solution
       $CuSO_4(aq)$ + $Fe(s)$
   (g) When copper carbonate is heated
       $CuCO_3(s)$ (heat)
   (h) When manganese oxide is added to hydrogen peroxide
       $H_2O_2(aq)$ (MnO_2)

2. (a) Write an equation for the reaction between solutions of silver nitrate and sodium chloride, describe what would be observed, and explain how this reaction is used.
   (b) Explain how barium chloride may be used to test for sulphates and illustrate with an appropriate equation.

3. (a) Write ionic equations for a neutralisation reaction and a precipitation reaction.
   (b) Write an ionic equation for zinc displacing copper from copper chloride solution and use it to explain why this may also be referred to as a redox reaction.

4. Name each of the following compounds and calculate its formula mass:
   MgO, Al_2O_3, KOH, $CuSO_4$, K_2CO_3, NH_4NO_3, $Al(OH)_3$, $Mg(NO_3)_2$.

5. Calculate the percentage of the named element in each of the following:
   (a) Lead in PbO, PbO_2, Pb_3O_4
   (b) Nitrogen in NH_3, KNO_3, $CO(NH_2)_2$, $(NH_4)_2SO_4$.

6. Calculate the mass of named substance involved in each of the reactions:
   (a) Sulphur needed to react with 8 g of oxygen to form sulphur dioxide
   (b) Calcium carbonate which on heating should give 70 g of calcium oxide
   (c) Iron(III) oxide, Fe_2O_3, which when reduced by carbon monoxide gives 2 kg of iron.

7. (a) Draw reaction pathway diagrams for reactions which are exothermic and endothermic.
   (b) Hydrogen reacts with oxygen $2H_2 + O_2 \rightarrow 2H_2O$. Use the bond energies given to calculate the overall energy change and determine if the reaction is exothermic or endothermic. (Energies, kJ/mol: O=O = 498; H–H = 436; H–O = 464)

8. The diagram shows the volume of carbon dioxide produced in a reaction between medium-sized marble chips and dilute hydrochloric acid, carried out at 20°C. Copy the graph and on the same axes show the curves you would expect to obtain if:
   (a) medium-sized pieces were used and the reaction was carried out at 40°C
   (b) larger pieces were used and the reaction was carried out at 20°C
   (c) small pieces were used and the reaction was carried out at 40°C.
   Label the curves carefully.

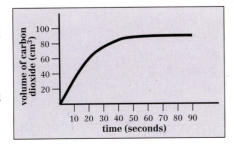

(d) Explain why each of the following has an effect on the rate of the reaction:
    (i) the size of the marble pieces
    (ii) the temperature at which the reaction is carried out.
(e) Draw and label a diagram of apparatus which would be suitable for this experiment.
(f) What do you understand by the term 'a fair test'?
(g) If you were investigating the effect of temperature on the rate of this reaction, which factors would need to be kept constant in order to make this a fair test?
(h) Give a word equation and a balanced symbol equation for this reaction.

9. Enzymes are described as biological catalysts.
   (a) What is a catalyst?
   (b) Name three enzymes and the reactions they catalyse.
   (c) Name two chemical catalysts and the reactions they catalyse.
   (d) Give one similarity between biological and chemical catalysts.
   (e) What is the difference between the effect of temperature on a biological and a chemical catalyst?
   (f) Describe the role of enzymes in the production of each of the following:
       (i) cheese   (ii) beer
       (iii) glucose syrup   (iv) medicine
       (v) biological detergents (vi) soft-centred chocolates.
   (g) Give two advantages of using biological detergents.
   (h) Study page 123 and give one disadvantage of using biological detergents.

10. Sulphuric acid is produced by an industrial process called the Contact Process. A flow diagram for the process is shown below. Study it and then answer the questions.

> **Stage 1** – Sulphur is burned in air.
>
> sulphur + oxygen → sulphur dioxide
>
> $S(s)$ + $O_2(g)$ → $SO_2(g)$

> **Stage 2** – Sulphur dioxide is mixed with more air and passed over beds of heated catalyst (vanadium(V) oxide) at 450°C.
>
> sulphur + oxygen ⇌ sulphur
> dioxide   trioxide
>
> $2SO_2(g)$ + $O_2(g)$ ⇌ $2SO_3(g)$

> **Stage 3** – Sulphur trioxide is first dissolved in concentrated sulphuric acid to give a thick fuming liquid called oleum. This is then diluted to give sulphuric acid.
>
> sulphur + water → sulphuric
> trioxide   acid
>
> $SO_3(g)$ + $H_2O(l)$ → $H_2SO_4(l)$

(a) What are the raw materials for this process?
(b) The reaction to produce sulphur trioxide is reversible. How can you tell this from the equation?
(c) A temperature of 450°C is used in Stage 2. Does this suggest that the forward reaction is endothermic or exothermic? Explain your answer.
(d) After studying the conditions for the Haber Process, you should understand that increasing the pressure will increase the yield of sulphur trioxide in Stage 2.
    (i) Explain why is this is the case.
    (ii) Why do you think that a pressure of only 2 atmospheres is used?
(e) Sulphuric acid and nitric acid are used in the production of many important and useful chemicals. List ways in which large industrial plants might cause environmental problems.

121

DEFINITIONS

Acid – A compound which dissolves in water to produce hydrogen ions

Activation energy – The minimum amount of energy particles must have in order to react

Alkali – A base which dissolves in water to produce hydroxide ions

Alkanes – A family of hydrocarbons the members of which only have carbon–carbon single bonds

Alkenes – A family of hydrocarbons the members of which have one carbon–carbon double bond

Alloy – A mixture of a metal with another element or elements, usually another metal

Anions – Negative ions which move to the anode during electrolysis

Anode – The positive electrode in an electrolysis

Atom – The smallest particle of an element which can take part in a chemical reaction

Atomic number – The number of protons in the nucleus of an atom

Base – A substance which reacts with an acid to produce a salt and water

Catalyst – A substance that increases the rate of a reaction but is not used up during the reaction

Cathode – The negative electrode in an electrolysis

Cations – Positive ions which move to the cathode during electrolysis

Chemical change – A change which involves the formation of new chemical substances

Chemical properties – Those properties of a substance which involve chemical change

Chromatography – A process for separating mixtures of coloured substances

Combination – A reaction in which two or more elements combine to form a compound

Compound – A substance made of two or more elements chemically combined

Condense – A change of state from gas to liquid

Covalent bond – A chemical bond in which two electrons are shared

Cracking – A process of breaking long-chain hydrocarbons into smaller molecules

Crude oil – A naturally occurring mixture of hydrocarbons

Decomposition – Breaking down a compound into simpler substances

Denaturation – Alteration of the shape of an enzyme, e.g. by heat, so that it will no longer work

Diatomic – Molecules containing two atoms

Diffusion – The ability of gases to spread out

Displacement – A reaction in which an element replaces a less reactive element in a solution of one of its salts

Distillation – A process for obtaining liquids from solutions by evaporation and condensation

Electric current – The movement of electrical charge

Electrolysis – The passage of electric current through a compound causing chemical reactions to occur

Electrolyte – The liquid through which current flows during electrolysis

Electron – A negatively charged particle which orbits the nucleus of an atom

Elements – (a) Chemicals which cannot be broken down into simpler substances

– (b) Chemicals which contain only one type of atom

Endothermic – A reaction which takes in heat from the surroundings

Enzyme – A biological catalyst produced by living organisms

Equilibrium – When the forward and reverse rates of a reversible reaction are the same

Erosion – A process, which involves movement, by which rocks are worn away

Eutrophication – When a lake or river is so rich in nutrients that too many plants grow

Evaporate – Change of state from liquid to gas

Exothermic – A reaction which gives out heat to the surroundings

Extrusive rock – Igneous rock formed when lava cools and solidifies as magma reaches the Earth's surface

Fermentation – An anaerobic respiration by which yeast uses sugar for energy without using oxygen

Filtrate – The liquid which passes through a filter paper during filtration

Filtration – A means of separating an insoluble solid from a liquid

Fossil fuels – Fuels such as coal, oil, gas, which have formed from the remains of plants or animals

Fractional distillation – The process of separating mixtures of liquids according to their boiling points

Fractions – Liquids separated from crude oil by fractional distillation

Freeze – Change of state from liquid to solid

Geological change – A change in which the structure or composition of rocks is altered

Giant structure – A structure of very many particles bonded together

Greenhouse gases – Gases present in the atmosphere which cause an increase in the Earth's average temperature

Group – A vertical column in the Periodic Table

Hydrocarbons – Compounds containing hydrogen and carbon as the only elements

Igneous rocks – Rocks formed from molten magma

Intrusive rock – Igneous rock formed when magma cools inside the Earth's crust

Neutralisation – a reaction between an acid and a base to give a salt and water

Oxidation – A reaction in which oxygen is gained or electrons are lost or hydrogen is lost

pH – A measure of acidity according to the concentration of hydrogen ions

Photosynthesis – The production of glucose from carbon dioxide and water by green plants using light

Physical change – A change in which no new substances are produced

Physical properties – Those properties of a substance which do not involve chemical change

Physical state – One of the states of matter

Plates – Large sections of the Earth's lithosphere

Polymerisation – A reaction in which molecules containing very long carbon chains are made from smaller units

Polymers – Compounds, made from smaller molecules, which contain very long chains of carbon atoms

Precipitate – A solid formed from solution as a result of two soluble substances reacting to form an insoluble one

Precipitation – A reaction which produces a precipitate

Products – Substances formed during a chemical reaction

Proton – A positively charged particle found in the nucleus of an atom

Radical – An ion which is made up of more than one atom

Reactants – The chemicals present at the start of a chemical reaction

Reduction – A reaction in which oxygen is lost or electrons are gained or hydrogen is gained

Relative atomic mass – The average mass of the atoms in an element using a scale on which the mass of carbon is twelve

Residue – The substance remaining in the filter paper after filtration

Respiration – A process of releasing energy from glucose

Reversible reaction – A chemical reaction in which the products can react to re-form the starting substances according to the prevailing conditions

Rocks – Mixtures of minerals which are present in the Earth's crust

Rusting – The corrosion of iron (or steel) which requires both air (oxygen) and water

Sacrificial protection – A method of rust prevention by using a more reactive metal which corrodes in place of the iron

Salt – A compound formed when the hydrogen of an acid is replaced by a metal

Sediment – Fragments of rock which have moved and settled

Sedimentary rocks – Rocks formed from layers of sediment

Solute – A solid which has dissolved to form a solution

Solution – A mixture in which one substance has dissolved in another

Solvent – A liquid which has dissolved another substance to make a solution

States of matter – Solid, liquid, gas

Substrate – A reactant which takes part in a reaction catalysed by an enzyme

Weathering – A process by which rocks are broken down into fragments *in situ*

FORMULAE

Aluminium oxide	Al_2O_3	Ethene	C_2H_4	Ozone	O_3
Ammonia	NH_3	Hydrochloric acid	HCl	Potassium carbonate	K_2CO_3
Calcium carbonate	$CaCO_3$	Hydrogen	H_2	Potassium chloride	KCl
Calcium chloride	$CaCl_2$	Hydrogen peroxide	H_2O_2	Potassium hydroxide	KOH
Calcium hydroxide	$Ca(OH)_2$	Iron oxide (Iron(III) oxide)	Fe_2O_3	Potassium oxide	K_2O
Calcium oxide	CaO	Iron(II) sulphate	$FeSO_4$	Propane	C_3H_8
Carbon dioxide	CO_2	Iron(III) sulphate	$Fe_2(SO_4)_3$	Propene	C_3H_6
Carbon monoxide	CO	Lithium chloride	$LiCl$	Silver nitrate	$AgNO_3$
Carbonic acid	H_2CO_3	Lithium hydroxide	$LiOH$	Sodium carbonate	Na_2CO_3
Chlorine	Cl_2	Lithium oxide	Li_2O	Sodium chloride	$NaCl$
Copper carbonate (Copper(II) carbonate)	$CuCO_3$	Magnesium carbonate	$MgCO_3$	Sodium hydroxide	$NaOH$
Copper chloride (Copper(II) chloride)	$CuCl_2$	Magnesium chloride	$MgCl_2$	Sodium oxide	Na_2O
		Magnesium nitrate	$Mg(NO_3)_2$	Sodium sulphate	Na_2SO_4
Copper oxide (Copper(II) oxide)	CuO	Magnesium oxide	MgO	Sulphur dioxide	SO_2
		Magnesium sulphate	$MgSO_4$	Sulphuric acid	H_2SO_4
Copper nitrate (Copper(II) nitrate)	$Cu(NO_3)_2$	Methane	CH_4	Sulphurous acid	H_2SO_3
		Nitric acid	HNO_3	Water	H_2O
Copper sulphate (Copper(II) sulphate)	$CuSO_4$	Nitrogen	N_2	Zinc carbonate	$ZnCO_3$
		Oxygen	O_2	Zinc chloride	$ZnCl_2$
Ethane	C_2H_6			Zinc sulphate	$ZnSO_4$

APPENDIX 1   The Periodic Table of Elements

KEY

Mass number A

Atomic number (Proton number) Z

| 1 | H Hydrogen 1 |

Group	Element

Group 1	Group 2											Group 3	Group 4	Group 5	Group 6	Group 7	Group 0
																	4 He Helium 2
7 Li Lithium 3	9 Be Beryllium 4											11 B Boron 5	12 C Carbon 6	14 N Nitrogen 7	16 O Oxygen 8	19 F Fluorine 9	20 Ne Neon 10
23 Na Sodium 11	24 Mg Magnesium 12											27 Al Aluminium 13	28 Si Silicon 14	31 P Phosphorus 15	32 S Sulphur 16	35 Cl Chlorine 17	40 Ar Argon 18
39 K Potassium 19	40 Ca Calcium 20	45 Sc Scandium 21	48 Ti Titanium 22	51 V Vanadium 23	52 Cr Chromium 24	55 Mn Manganese 25	56 Fe Iron 26	59 Co Cobalt 27	59 Ni Nickel 28	64 Cu Copper 29	65 Zn Zinc 30	70 Ga Gallium 31	73 Ge Germanium 32	75 As Arsenic 33	79 Se Selenium 34	80 Br Bromine 35	84 Kr Krypton 36
85 Rb Rubidium 37	88 Sr Strontium 38	89 Y Yttrium 39	91 Zr Zirconium 40	93 Nb Niobium 41	96 Mo Molybdenum 42	99 Tc Technetium 43	101 Ru Ruthenium 44	103 Rh Rhodium 45	106 Pd Palladium 46	108 Ag Silver 47	112 Cd Cadmium 48	115 In Indium 49	119 Sn Tin 50	122 Sb Antimony 51	128 Te Tellurium 52	127 I Iodine 53	131 Xe Xenon 54
133 Cs Caesium 55	137 Ba Barium 56	139 La Lanthanum 57	178 Hf Hafnium 72	181 Ta Tantalum 73	184 W Tungsten 74	186 Re Rhenium 75	190 Os Osmium 76	192 Ir Iridium 77	195 Pt Platinum 78	197 Au Gold 79	201 Hg Mercury 80	204 Tl Thallium 81	207 Pb Lead 82	209 Bi Bismuth 83	210 Po Polonium 84	210 At Astatine 85	222 Rn Radon 86
223 Fr Francium 87	226 Ra Radium 88	227 Ac Actinium 89															

Elements 58–71 and 90–103 have been omitted.

The value used for mass number is normally that of the commonest isotope, e.g. ^{35}Cl not ^{37}Cl.

Bromine consists of approximately equal proportions of ^{79}Br and ^{81}Br

APPENDIX II   Materials and their uses

THE FRACTIONS FROM CRUDE OIL

Refinery gas	Liquefied by pressure, flammable	Bottled gas e.g. Calor gas
Gasoline/Naphtha	Volatile, flammable	*Petrol*, fuel for cars
Kerosene	Flammable	Jet fuel, *paraffin*
Diesel		Fuel for heavy vehicles and cars
Lubricating oil	Less flammable, viscous	Lubrication
Fuel oil	Special burners needed	Fuel for heating systems
Wax/grease	Difficult to ignite, viscous	Polishes, Vaseline
Residue	Very viscous	*Bitumen*/tar for roofs and roads

POLYMERS

Polymers/plastics are generally unreactive, resist corrosion, are cheap to produce, of low density, flexible, non-conducting, and waterproof. There is a large variety of these materials and similarly they have a wide range of properties and uses.

Poly(ethene)	*Low density*: tough, flexible, soft	Detergent bottles, carrier bags
	High density: stiffer, harder	Bowls, buckets
Poly(propene)	Light, hard, impact resistant	Crates, rope, chair seats
Poly(styrene)	*Expanded*: light, good insulator	Sound/heat insulation, packaging
Poly(vinyl chloride)	*uPVC*: tough, resists weathering	Pipes, gutters, window frames
	Plasticised: soft, flexible, insulator	Hosepipes, electrical insulation
PET	Strong, chemical resistant, clear	Drinks bottles

METALS

Are generally stronger but more dense than plastics. They are often used as alloys, which are stronger, harder, and less likely to corrode than pure metals.

Iron	Cheap, strong	To make steel, catalyst in the Haber process
Nickel	Unreactive	Alloys for coins, catalyst in margarine manufacture
Copper	Unreactive, good conductor	Water pipes, electric cables, windings in motors
Aluminium	Low density, shiny, reflects radiation, good conductor	Cooking foil, CDs, overhead power cables, alloys for aeroplane manufacture, drinks cans
Platinum	Unreactive	Catalyst for oxidising ammonia
Chromium	Reflective, unreactive	A component in stainless steel
Tungsten	Strong, high melting point	Tungsten steel
Titanium	Hard, strong	Titanium steel

NON-METALS

Hydrogen	Reacts with nitrogen	Manufacture of ammonia
	Hydrogenates vegetable oils	Manufacture of margarine
Carbon	As coke	Extraction of iron
Helium	Less dense than air, non-flammable	Airships, balloons
Neon	Emits colour with electric current	Advertising signs, neon lights
Argon	Colourless, unreactive	Replaces air in light bulbs to prevent filament burning
Krypton	Colourless, unreactive	Lasers
Chlorine	Poisonous, kills micro- organisms	To make bleaches, sterilising liquids
		Chlorination of drinking water to kill bacteria
		Manufacture of hydrochloric acid, PVC, CFCs
Iodine (solution)	Kills bacteria, reacts with starch	Antiseptic solution, testing for starch

COMPOUNDS

Ammonium nitrate	Nitrogen compound	Fertilizer
Ammonium sulphate	Nitrogen compound	Fertilizer
Calcium carbonate	Limestone	Manufacture of cement, glass, steel
Calcium hydroxide	Slaked lime	Neutralising acidity in lakes affected by acid rain
Calcium oxide	Quicklime	Neutralising acidity in soil
Silver halides	Darken in light	Photographic film and paper
Sodium carbonate		Manufacture of glass, washing soda
Sodium chlorate	In place of chlorine solutions	Bleach, sterilising liquids, water treatment
Sodium chloride	Solution can be electrolysed	Producing chlorine, hydrogen, sodium hydroxide
	Lowers freezing point of water	Treating icy roads
Sodium fluoride		In toothpaste
Sodium hydroxide		Manufacture of soap, paper, ceramics
Sulphuric acid		Manufacture of ammonium sulphate

APPENDIX III Chemical data

ATOMS
(At. No. is the Atomic Number) (Mass is the Relative Atomic Mass)

Element	Symbol	At. No.	Mass	Element	Symbol	At. No.	Mass
aluminium	Al	13	27	krypton	Kr	36	84
argon	Ar	18	40	lead	Pb	82	207
barium	Ba	56	137	lithium	Li	3	7
beryllium	Be	4	9	magnesium	Mg	12	24
boron	B	5	11	neon	Ne	10	20
bromine	Br	35	80	nitrogen	N	7	14
calcium	Ca	20	40	oxygen	O	8	16
carbon	C	6	12	phosphorus	P	15	31
chlorine	Cl	17	35.5	potassium	K	19	39
copper	Cu	29	64	silicon	Si	14	28
fluorine	F	9	19	silver	Ag	47	108
helium	He	2	4	sodium	Na	11	23
hydrogen	H	1	1	sulphur	S	16	32
iron	Fe	26	56	zinc	Zn	30	65

One mole of gas at normal temperature and pressure occupies 24dm^3

IONS
Positive ions

sodium	Na^+	magnesium	Mg^{2+}	aluminium	Al^{3+}
potassium	K^+	calcium	Ca^{2+}	iron(III)	Fe^{3+}
silver	Ag^+	zinc	Zn^{2+}		
ammonium	NH_4^+	copper	Cu^{2+}		
lithium	Li^+	lead	Pb^{2+}		
		iron(II)	Fe^{2+}		

Negative ions

chloride	Cl^-	oxide	O^{2-}	phosphate	PO_4^{3-}
bromide	Br^-	sulphide	S^{2-}		
iodide	I^-	sulphate	SO_4^{2-}		
fluoride	F^-	carbonate	CO_3^{2-}		
hydroxide	OH^-				
nitrate	NO_3^-				
hydrogencarbonate	HCO_3^-				

CHEMICAL TESTS

Oxygen	relights a glowing splint
Hydrogen	pops with a lighted splint
Carbon dioxide	turns limewater cloudy ('milky')
Chlorine	bleaches moist red litmus paper
Ammonia	turns moist red litmus paper blue
Alkenes	decolorise bromine water
Water	changes anhydrous copper sulphate from white to blue
Water	changes cobalt chloride from blue to pink
Chlorides (aq)	give a white precipitate with silver nitrate solution
Sulphates (aq)	give a white precipitate with barium chloride solution

SAFETY IN THE LABORATORY

HAZARD SYMBOLS

Symbol	Meaning	Precautions
Harmful (Xh)	May cause limited health risk if taken in by mouth or inhaled or if absorbed by the skin. **Examples** • copper carbonate • copper oxide	Do not breathe dust, spray or vapour. Avoid contact with the skin. Wash hands thoroughly before eating or drinking afterwards. If there is contact with the eyes, rinse immediately with plenty of water.
Irritant (Xi)	May cause soreness or irritation when in repeated or prolonged contact with the skin or if inhaled. **Examples** • enzymes, including biological detergents • moderately concentrated hydrochloric acid • moderately concentrated ammonia solution • lime water	Do not breathe dust, spray or vapour. If there is contact with the eyes, rinse immediately with plenty of water.
Toxic	May cause severe health risk or even death if taken in by mouth or inhaled or if absorbed by the skin. **Examples** • copper chloride (solid) • barium chloride • lead compounds (solids) • ozone	Carry out experiments in a fume cupboard. Wear suitable protective clothing covering eyes, face, and hands. If there is contact with the eyes or skin, rinse immediately with plenty of water. In case of accidents, or if you feel unwell, seek medical advice immediately.
Corrosive	May cause burns or destruction of living tissue on contact with the skin. **Examples** • concentrated hydrochloric acid • moderately concentrated sulphuric acid • sodium metal • calcium oxide	Wear suitable protective clothing covering eyes, face, and hands. Remove immediately all contaminated clothing. If there is contact with the skin, rinse immediately with plenty of water. If there is contact with the eyes, rinse. immediately (for 15 minutes) with water and seek medical advice.
Oxidising	Provides oxygen. May cause explosion or fire. **Examples** • concentrated nitric acid • hydrogen peroxide • potassium manganate(VII)	Read and use according to instructions. Containers should be stored in a cool well-ventilated place and kept tightly closed. Store away from sources of ignition and heat. Dispose of containers and oxidising substances safely.
Highly flammable	Ignites easily, flash point below 21°C. **Examples** • sodium metal • potassium metal • ethanol • petrol	Use only in flameproof areas. Store away from sources of ignition and heat. Do not breathe spray or vapour. Avoid static discharge.

SAFETY IN THE LABORATORY

1. You should not enter a laboratory without permission.
2. Follow instructions when using equipment or chemicals. Be aware of any hazard symbols on containers of chemicals.
3. Wear eye protection when carrying out experiments. Accidents involving chemical splashing may occur while apparatus is being cleared away.
4. When using a Bunsen burner, hair and loose clothing should be tied back.
5. When working with liquids it is safer to stand up so you can move out of the way quickly if there is a spillage.
6. Do not taste anything. If you get chemicals on your skin wash them off. You should wash your hands thoroughly after using chemicals.
7. Report accidents or breakages. This is so that injuries can be attended to and spilled chemicals or glass fragments can be dealt with safely.
8. Keep your work area safe, clean and tidy.

ASSESSING RISK – A CHECK LIST

1. What are the details of the experiment?
2. What are the hazards?
3. What might go wrong?
4. How serious would it be if something went wrong?
5. How can the risks be controlled?
6. What action would I need to take if something went wrong?

INDEX